Blood And Steel The Rise Of The House Of Krupp

Bernhard Menne

Menne, Bernhard.

Blood and steel; the rise of the house of 1938.

BLOOD AND STEEL

BLOOD AND STEEL

THE RISE OF THE HOUSE OF KRUPP

By

BERNHARD MENNE

NEW YORK
LEE FURMAN, INC.
1 9 3 8

Published in England under the title
KRUPP, THE LORDS OF ESSEN

Translated from the German by G. H. Smith

Printed in the United States of America

PREFACE

ZAHAROFF, the Man of Mystery, aroused the curiosity of a dozen of post-War sensation-seeking publicists. But the House of Krupp, which could boast the friendship of two German Emperors, its guns making the background for a half-century of mad armament policy, and again doing so by the will of Krupp's new patron, Adolf Hitler, has never yet, throughout the many generations of its existence, been the subject of a genuinely critical history.

The fact, though surprising, is by no means accidental. For this firm, surpassing in adroitness of intelligence all its competitors in the armament industry, has successfully camouflaged its activities in a rather novel manner. Several hundred books and pamphlets and more than a thousand scientific papers seem to imply a veritable publicity-mania, which, however, is a stroke of true genius: i.e., concealment of the truth behind a mass of cleverly dished-up data.

The author confesses that herein lay the real difficulty of his task. Here was no charitable obscurity in which every ascertained fact could prove a welcome guide to the Unknown. Step by step mountains of paper had to be pushed aside in order that what was essential could be found in spite of thousands of printed pages. And utmost care was taken to investigate and weigh the sources of all information. Definite, undisputed statements alone could be utilised here. The author, lacking all talent for fiction, did not allow his imagination play in any single instance. He himself retreated behind the facts presented.

It is obvious that the association of politics and business, steel and the destiny of nations, revealed in these pages is not to be considered peculiar to the history or the present condition of Germany. Wherever the name "Krupp" appears, let the Frenchman substitute "Schneider"; the Englishman, "Vickers"; and any other country, its corresponding firm.

Not out of blind antagonism nor hatred has this criticism of one phase of German actuality been born. Rather out of sadness; sadness for a kindly but misguided people; for a noble country; and especially for one strip of earth beside the Ruhr which will always remain the Promised Land to the writer.

CONTENTS

PART V

THE KRUPP A.-G—THE LIMITED COMPANY

PART I

THE ANCESTORS

Arndt Krupe—The Arms Dealer—The Three Brothers—The First Ironworks

THE ANCESTORS

ARNDT KRUPE

IN 1599, the year of the Plague, a small German city near the right bank of the Rhine staggered through the horror in the grip of the Black Death. Flourishing families were wiped out overnight; entire streets, the chroniclers relate, looked like graveyards "in their sad desertion."

In the midst of this distracted heap of humanity the wine merchant on Salt Market kept his head. Property and money had lost their meaning for most of those who survived, but he made the best of death and purchased for a few thalers extensive "gardens and pastures" before the city gates.

That was the ancestor of the House of Krupp.

The register of the merchants' guild of Essende for 1587 refers to a new member, one Arndt Krupe. No further particulars are given and nothing is said of his origin, his age, or when he arrived. Writers of Krupp history have vainly sought to establish his origin. There is reason to believe that he hailed from Ahrgau, on the left bank of the Rhine, perhaps from Ahrweiler where the names Krupe, Kruyp, and Krup were fairly common at that time. At the confluence of the Ahr and the Rhine, in the present village of Kripp, is a "Krippe" or "Kruppe," *i.e.*, a crib, which may have been inhabited by one of the forefathers as ferryman—but these are mere conjectures.

However dim his past may have been, the man quickly made his mark in his new home and the city records of Essende of that period refer continually to him. They tell us that he was a merchant, dealing in wines, spirits and occasionally cattle, who soon took up the trade in Dutch "spices," groceries. This display of manifold business activity by a newcomer is not exceptional, as acquisitiveness is a common feature of new settlers and immigrants. It is worthy of note that Krupe was a disciple of the Reformation. If he came from the Valley of the Ahr—the Abbess owned vineyards there and he began as wine dealer—his immigration to the Protestant stronghold of Essende may have been due to religious persecution.

Arndt Krupe must have been considered a man of ability as he was entrusted with public business. When differences arose between the Reformed Party and the Lutherans, the magistrates rallied to the "Lords Protector" of the city, the Counts von Neuburg and Brandenburg: "On the 7th April, a licentiate, Master Basselrodt, and a gentleman, Master Arndt Krup, delivered a reply in writing from the Worshipful Council to the Noble Lords." There is nothing surprising in the fact that this may refer to a religious matter. Krupe was an active member of the Lutheran congregation, which was using a hymnal published by Arnold Krupe and Tieleman Leimgarten.

Arndt Krupe died in 1624. In the thirty-seven years since he emerged from obscurity he had won for himself and his children, Anton, Catharina, Georg and Margarethe, the position of respected citizens. One of the last items of information concerning him is as hard and close-fisted as he was himself; for the sum of two rats-thalers, so the records state, he bought a gravestone from the public quarry. It was for himself.

THE ARMS DEALER

Essende's trade in armaments was originally in the hands of the master armourers. They made firearms in small workshops with the assistance of journeymen and apprentices and sold them to the agents of armies and princes. Their profits as well as their professional arrogance aroused the envy of the rich mercantile families to whom the Council had given dangerous power. An edict issued by the magistrates permitted the armourers to sell their wares within city precincts only, while the wholesale export trade was reserved for the merchants. This, in the case of a production dependent upon export, was a deathblow to the independence of the crafts guilds; they came under control of the Council, which appointed a master gunsmith to supervise manufacture.

Among the names frequently appearing in connection with the arms trade of Essende at that period is that of the Krösens, an old patrician family which had provided several city fathers. It was, therefore, a great honour for the merchant Krupe that his eldest son Anton—born in 1588—should wed Gertrud Krösen. Anton Krup—as his name now appears in contemporary records—was in his father's business. He dealt in spirits, Spanish wines and groceries, and paid taxes on "imports and exports." Something of the restless spirit characteristic of later Krupps appeared in him, for he soon turned to the business which promised high profits, despite the speculative nature of its basis, *i.e.*, the trade in armaments.

His name first appears in connection with the arms trade in 1615; shortly after that date the great war broke out. For thirty years the Empire was the battleground of foreign armies; burned houses, ruined crops, and plundered moneybags marked their passage. The palace of the Abbess was presided over by Maria Klara von Spaur, in whom burned the inquisitorial zeal of the Counter-Reformation. She denounced Essende to the Spaniards as "a fanatically evangelical and bitterly hostile city, whose bringing back to obedience will be a meritorious work." From one of her trips to Brussels she brought back five companies of Italians, who imposed a short-lived council on the city. Billetings, not far removed from sackings, seemed to go on forever, and the despairing citizens of those years petitioned the Emperor "for leave to emigrate to the Electorate of Cologne."

One industry prospered—in 1608 there were twenty-four gunsmiths and dealers engaged in the firearms trade, in 1620 there were already fifty-four. The site of Essende, between two contending parties, made for good business. It sold to the Protestant Netherlands, the Catholic Electorate of Cologne, and neutral Reformed Brandenburg. Anton Krup threw all his energy into the industry of war—the first of the House of Krupp to do so! His turnover reached one thousand gun barrels—quite a good proportion of the total annual output of some fifteen thousand. The Krup wine business flourished those days also, with so many thirsty soldier throats looking to the City Council for assuagement. No one wondered that the wine merchant and arms exporter

Krup could buy a house on the Rott and that he leased the right to tax flour and grain.

Anton Krup had not only taken over his father's business, he took over as well the various public offices held by his sire. He was, apparently, a good negotiator, not devoid of legal knowledge, was frequently named among delegates deputed to settle disputes with leaders of billeted troops. In 1641 the council issued a written authority for "our highly honoured patriot, the nobly born Mr. Anthon Krupp" to Frederick William of Brandenburg, the "Great Elector" of Prussian history. The city petitioned its Lord Protector that it might be taken up into the neutral zone of Brandenburg's West-German possessions. Although we do not know whether the petition was presented in Berlin or in Cleves, the Elector's reply was, at all events, a favourable one.

There came, however, a change in the fortunes of this successful business man when he reached the age of fifty years. Anton Krup's fortune, unlike that of his father, was not invested in land and rents and it shrank visibly. There may have been losses in the gun business, but it is more probable that certain peculiarities in the character of the man himself became more pronounced with advancing years. Even in those stormy days he seemed to have been particularly truculent and litigious. His wife's affairs—chiefly matters connected with legacies—brought his name before the courts and it appeared in the legal records of those days more frequently than any other. In keeping with the spirit of the times the litigants laid violent hands on one another and it is recorded that Anton Krup "is fined 8 dollars for beating Dr. Hasselmann in the street."

THE THREE BROTHERS

In 1648 the bells of peace rang out in the Empire and the appalling Thirty Years' War came to an end. Essende reverted to its everyday avocations after buying off the last troops billeted in the city. The unending horrors must have produced openings for fresh and untapped sources of energy. A new abbess was appointed in the person of the still youthful Countess Anna Salome and at the same time the twenty-five-year-old town clerk, Matthias Krupp, signed his name under his first official document.

The young *secretarius* was the only child of the wine merchant Georg Krupp, second son of Arndt Krupe. The fate of his parents was a tragic one; "*obit peste*," is the grim entry in the city register of 1623. Two-year-old Matthias came under the guardianship of his uncles, Anton Krup and Matthias Klocke, both educated and cultured men. It was doubtless due to them that the boy, who came into a substantial inheritance, took an unusual course; he attended the Duisburg Grammar School and then went to the "Gymnasium Illustre" in Bremen, which was equivalent to a university. He was intended for an important office, that of *secretarius* or town clerk. While the mayor and city council changed frequently, the town clerk was appointed for life and combined in his person permanency with administrative knowledge. He enjoyed extensive powers, was authorized to make payments on his own responsibility, and made up the accounts for taxes and disbursements. His official salary was negligible, but fees and numerous honorariums made life in the important office bearable.

The private life of Matthias Krupp was completely eclipsed by his official one. He married at an advanced age and inherited the imposing "Haus Zur Krone," which had seen the Elector of Brandenburg as guest within its walls. He added to his inherited property within the city walls by purchasing large tracts of land on the north-western boundaries of the city, which were to become the site of the steel foundry centuries later.

The official obituary notice in the city records is significant: "Secretarius Matthias Krupp died on the 8th February, 1673. G. Krupp, his son, succeeded him, Dr. Westerdorff acting, until he took up office." The eldest of the three sons was only sixteen years old, so that his father's post had to be kept vacant for him. The municipal office became a heritage of the Krupp family, and from being obscure newcomers they had now attained the position of recognized patricians.

Georg Dietrich, the eldest son, born in 1657, succeeded his father while still a boy. He went through a brief course of study at Duisburg while his father's old friend Westerdorff kept the billet open for him. Georg Dietrich took over the office of town clerk at the age of twenty-one and filled it for sixty-four years.

Sixty-four years . . . the century of sanguinary religious wars passed, in the east the military power of Prussia began to rise, while across the Rhine shone the glamour of the "Roi Soleil." The gladly welcomed "Sæculum Humanum" opened with the World War of the Spanish Succession and then came the first Frederican attack on the Empire. During all this super-humanly long period, Georg Dietrich sealed the municipal records of Essende and steadily built up the family fortune and power.

Apart from the political and intellectual activities of his office, Georg Dietrich displayed an astounding energy in business. It is hard for anyone making a study of its nature and extent to realize that this business was only a side line, as it was so vast. He purchased houses, land, and gardens within and without the city gates. Besides which were financial transactions, money lending —down to quite small sums—in the regular way of business. He exacted mortgages on houses for loans of a few dollars and acquired one house from an aged couple who had fallen upon evil days and become chargeable to the poor-law authorities. Georg Dietrich bought the house from the guileless old couple for a trifling amount, and added to it a second one by foreclosing on the mortgage he held on it. He rented these houses to the poorest citizens, a business that demanded callousness but yielded good profits.

Georg Dietrich's brothers also occupied important positions. The youngest, Arnold Krupp, had been awarded a Doctor of Laws degree in Giessen for a Latin thesis on feudal law. His writings might lead one to assume that he became a distinguished man of letters, but his later life does not appear to have had any connection with literary matters. Apart from ordinay financial transactions, he collected rents from tenants of the Count of Styrum. As the junior member of the council, he was, at the age of thirty, elected mayor, in which capacity he neglected no opportunity to further his own interests.

Beside the businesslike Arnold and the lordly Georg Dietrich, the third brother, Matthias, cut an inconspicuous figure. He followed the calling of a cloth and wool merchant—an astonishing profession for a Krupp of that period to engage in. His property and the extent of his business were small and although he farmed the taxes on corn and flour and occupied the lucrative position as superintendent of the orphanage, his drawings from city corporation funds appear to have been unusually modest.

Georg Dietrich was now an old man of over eighty, surrounded by elderly people who were born when he was already in office. It was time to nominate a successor. His own son had died in childhood, but the office inherited from his father had to pass to another Krupp. None of the three sons of his brother Arnold had shown any great ability, but he finally selected the second eldest, Henrich Wilhelm, to be *secretarius adjunctus.*

Georg Dietrich died on March 2nd, 1742, at the age of eighty-five and with his passing the great days of the Krupp family came to an end. In 1749 they celebrated the centenary of the holding of the office of town clerk by a member of the Krupp family, but it was a last flicker, as the actual office-holder, Henrich Wilhelm Krupp, was a weak and inefficient man. He unsuccessfully turned his attention to mining; the old Essende colliery, named the Secretarius Pit, is said to have been called after him. But coal mining was still too precarious an industry to prove remunerative and Henrich Wilhelm was finally forced to sell his house to pay his more pressing creditors. He died in 1760, the last town clerk to bear the name of Krupp. His only son soon followed him and his affairs were left in such disorder as to make it necessary for his widow to file her petition in bankruptcy. When she left the city, the creditors stated that "she secretly fled."

THE FIRST IRONWORKS

The generation following that of the three brothers saw the family alarmingly diminished. Georg Dietrich's only son died in childhood and of the sons of Dr. Arnold only the eldest, Jodokus, remained. The continued existence of the family was dependent on him.

Friedrich Jodokus Krupp, born in 1706, was the link between the medieval greatness and the industrial rise of the Krupps. In him the family commercial instincts rebelled against the patrician outlook of the preceding generation and he shook off the trammels of tradition to face the realities of everyday life. Being a merchant, he first tried to deal in cattle and then to establish a business for groceries and spices. Aged barely twenty, he married a wealthy heiress of thirty-two summers. Their union was childless, but material blessings were plentiful. In 1737 he bought a house in the centre of Essende at the corner of the Flax Market and the Limbeckerstrasse. This house became the ancestral home of the junior branch of the Krupp family, but Jodokus and his wife were fated to occupy it for a brief period only. The wife died and left him a middle-aged widower, presumably the last of his race. He developed an interest in public affairs and was elected to the city council, where he occupied several minor honorary posts.

Then unexpectedly, Jodokus, in his late forties, married the attractive and singularly intelligent Helene Amalie Ascherfeld, nineteen-year-old daughter of a neighbour and contemporary. She brought fresh energy and business acumen to the decaying family and it is interesting to note that she also was descended from

Arndt Krupe whose daughter Margarethe was her father's great-grandmother.

The young wife took over the direction of the business, in which she displayed great ability, while Jodokus devoted his time to public life, playing a respected but inconspicuous part. He died at the age of fifty-one and Helene Amalie continued the business under the name of "Widow Krupp."

This woman completely overshadowed her son Friedrich Wilhelm Krupp, who inherited his father's intellectual modesty. He preferred social activities, became a lieutenant in the rifle association, and a member of the municipal council. His only claim to distinction was as a father. His marriage to Petronella Forsthof, the heiress of a yeoman from near Düsseldorf and the child of a sixteen-year-old father and an eighteen-year-old mother, brought a lighter and more adventurous strain into the bourgeois character of the Krupp family.

Friedrich Wilhelm acted as accountant to his mother's business and compiled an inventory of the family possessions which sheds an interesting sidelight on the indebtedness of the peasantry to the city merchants of those days. The total extent of the Krupp fortune amounted to over 120,000 thalers. The accounts of the firm and the various business trips made on its behalf appear to have exhausted the energies of young Krupp, who died in 1795, leaving three children, Helene, Friedrich and Wilhelm.

The death of her son did not lame Helene Amalie's energy. The older she grew, the more daring grew her business activities. For over two thousand thalers she purchased the "Walkmuhle," the fulling mill just north of Essen, and therewith entered the field of industry, which was later to determine the destiny of the House of Krupp.

Eberhard Pfandhöfer founded the small "Good Hope" Ironworks, at Sterkrade, on the borders of Prussia and Cleves, in 1781. Pfandhöfer was a sound ironmaster but he lacked the commercial instinct and was short of money. The Krupps frequently helped him with substantial loans, but it soon became evident that the works were too small to pay their way. Their subsequent fate is revealed by the following entry in Helene Amalie's inventory: "The Good Hope Ironworks are situated in Starkrad and were the property of Eberhard Pfandhöfer. As he fled privily, his estate went into bankruptcy, and due to my heavy claims on it I was forced to buy it in at the public auction, paying 12,000 thalers, that is, 15,000 thalers Berlin currency, for all the buildings, plant, rights, and goodwill." This keen business woman put in her own manager at first and attempted to run the place herself, but when she selected her young grandson Friedrich to act for her the decisive move in the history of the firm of Krupp was made.

Six generations had followed the passing of Arndt Krupe and the family was now represented by the great-grandson of his great-grandson. The earliest record of the founder of the family appeared in 1587 and Friedrich Krupp was born in 1787. The two intervening centuries reveal the interesting fact that none of the family was a farmer or an artisan—they were all traders, money-lenders, and office-seekers, men whose thoughts centred on money and who heaped profits with skilled hands. Those that came after did not belie their ancestors.

When the hour struck, when the Cyclops of Industry began to build up its work, the Krupps took hold. The trader's blood in their veins gave them shrewd perception to see and seize the great chance.

PART II

THE FOUNDER

Smuggler and Speculator—The Bankrupt

THE FOUNDER

SMUGGLER AND SPECULATOR

PETER FRIEDRICH, grandson of Helene Amalie Krupp, was born July 17th, 1787, and with his birth began the true industrial history of the Krupps. That was the year France declared national bankruptcy, thereby starting an avalanche of far-reaching events. Frederick II of Prussia had just died. The city was the scene of a final conflict between the municipal council and the Lady Abbess Maria Kunigunde who owed her office to a fruitless bridal night with the son of Maria Theresa. The citizens filled in a colliery shaft belonging to the Abbess whose armed retainers forced their way into the city to wreak vengeance. Then came the troops of the Allied Kings, their retirement being followed by the entry of the French armies. In 1802 Prussia seized Essen and the thousand years' independence of the Imperial Abbey territory came to an end. A few more years saw the defeat of Prussia and Essen passed under French rule.

It was a time in which youth came to realize that the only permanent thing was change. Friedrich Krupp, who lost his father at the age of eight, was ever after to bear the mark of these unsettled and eventful days. Other influences were negligible. The College of Essen was at its lowest ebb at the close of the century and Friedrich's education was faulty, despite several years spent in it. He could scarcely write a letter grammatically, and was an unstable, impetuous young man, disinclined to concentrate on hard work.

When Essen had become Prussian the grandmother attempted to extend the Gute Hoffnung Ironworks and asked the government authorities in Berlin for a subsidy, which she received in the shape of orders—including orders for solid shot. This was the second instance of armaments dealt in by Krupp since the small arm trade of the Thirty Years' War. Helene Amalie evidently contemplated leaving the ironworks to her grandson when she acquired them. After his brief period of business training in the Flax Market store she transferred the management of the ironworks to him in 1807 and soon afterwards assigned the entire property "to provide him with a proper income." The grandmother's action in providing for her nineteen-year-old grandson was doubtless inspired by his engagement to the sixteen-year-old Therese Wilhelmi.

Young Krupp was not qualified to manage the ironworks, and his letters of that period reveal him as a truculent and ignorant youth whose first action was to dismiss the experienced works manager. His stay at the works was chiefly noteworthy for bringing him into contact with two men, who were among the pioneers of Rhenish-Westphalian industry: Dinnendahl and Jacoby. The former swineherd and joiner Franz Dinnendahl made his mark by his mechanical genius. At a time when, according to himself, the Ruhr territory could not produce a millwright capable of making a decent screw, he built the first steam engines. Gottlob Jacoby of the neighbouring works was a highly trained ironmaster, expert in metallurgy, and at the moment investigating the casting of steel.

His first few months at the ironworks proved the young man a dangerously unstable character. Dinnendahl, likewise lacking in business acumen, advised him to alter the nature of the entire output. The works had hitherto made stoves, plates, pans, pots, and weights, all highly remunerative lines. He now proposed to embark on the expensive and financially still doubtful manufacture of engine components, pistons, cylinders, and steampipes. He also planned to put up new buildings, but failed to carry out this scheme, as the alarming reduction in the turnover upset his grandmother. Taking advantage of the grandson's illness, she deprived him of the ownership of the works and put them up for sale. There was no lack of prospective purchasers, as some of the leading founders of the Rhenish-Westphalian heavy industry wanted to amalgamate the Neuessen Works near Essen and the Cologne Antony Works with the Cleves Gute Hoffnung Works. This syndicate included Gottlob Jacoby, Heinrich Huyssen and the latter's sons-in-law Franz and Gerhard Haniel, wealthy colliery proprietors. The sale took place in November, 1808, at the relatively high figure of 37,800 thalers.

Friedrich Krupp, now married, returned to commerce and took up his abode in his grandmother's house. Wilhelm and Helene (who had married Second Lieutenant von Müller) remained with their mother. But Friedrich soon tired of shopkeeping and turned his attention to an enterprise promising an easier way of making money.

We know from the chronicles of the Rothschild family how the shortage of manufactured goods brought about by Napoleon's "Continental embargo" fascinated the mercantile community. Nathan Rothschild went to London, and with the textiles and bar gold imported by his mysterious methods laid the foundations of the future greatness of the Frankfort banking house. In 1809 the blockade was tightened by the prohibition of import of food-stuffs from the Netherlands to the Rhenish States.

In a decree published at Schönbrunn Napoleon ordered the establishment of a chain of customs posts extending from Rees to Bremen. Intensive smuggling to the Westphalian and Rhenish cities began, and this was just the kind of undertaking that appealed to young Krupp rather than honest hard work. In November, 1809, he made an agreement with the mercantile firm of Winters, Mensinck & Co., in Borken, concerning the smuggling of Dutch colonial produce, mainly coffee, indigo, and sugar. Transit from Amsterdam to Essen was at "joint risk." Krupp advanced his Borken friends about 10,000 thalers which he raised with the help of his family.

It soon became obvious that the "risk" was a real one, as smuggling was extremely dangerous. The blockade was intensified and led to confiscations and shooting (with casualties), but young Krupp stood to lose only a few hundred dollars on each "risk" and continued his dangerous trade quite unconcernedly, extending the zone of his operations as far as Frankfort. His friends' letters from Borken became more and more pessimistic until they finally announced that "all means of passing the goods through have now been stopped." They advised that no further orders be placed in Amsterdam and the lucrative trade came to an end.

Meanwhile there had been a great change in the Flax Market shop; grandmother Helene Amalie died March 9th, 1810. This remarkable woman, born in the early years of the preceding century, lived to see the dawning of a new age. After a widowhood of fifty-five years she left a substantial fortune, mainly in real estate and well-secured mortgages, apart from some 120,000 thalers in cash. Friedrich Krupp inherited an initial legacy of 40,000 thalers which may be regarded as the financial foundation of his future business. He also inherited the store in the Flax Market to which he devoted some attention with a view to converting it into a wholesale business in coffee and sugar.

The Western districts of the Empire had begun to show economic progress since the middle of the eighteenth century. Mineral wealth was ample for building up heavy industries and deposits of iron ore abounded in many districts, although it was mined in primitive galleries and brought out by the sackload, usually on human shoulders. Smelting was carried out generally in the mountains near the charcoal kilns, as charcoal was needed to separate the metal from the ore in the little smelting furnaces. Power obtained from adjacent waterfalls was used to operate hammers for forging iron. The entire industry was dependent on natural resources which were liable to fail. An adequate technical solution of the problem was not easy to find.

There was, indeed, plenty of coal mined in the valleys of the Ruhr, the Wurm and the Saar, but it was the contemporaneous discovery of the uses of "coke" in England that enabled the production of iron to be carried on independently of wood supplies. Iron ore came back to the neighbourhood of the collieries—industry left the mountains and descended to the plains.

This fortuitous combination of iron and coal opened up new vistas. The dawning era of machinery demanded a raw material of great toughness, iron steel, but its production on a large scale lacked uniformity of material, which was imperative for modern machine construction. Cast steel was the first material to meet this requirement. A Sheffield clockmaker, Benjamin Huntsman, himself a user of steel, solved the problem of smelting steel without taking up carbon, the addition of which produced cast iron. The means employed were simple; small, covered crucibles, heated in forges, were raised to the requisite temperature and their use enabled steel of a hitherto unattainable quality to be produced. The process remained a British secret for half a century and led to the phenomenal rise of the little town of Sheffield.

Complete cessation of the import of British goods because of the Continental embargo, led to a fatal shortage of this essential raw material in the Ruhr District and threatened to extinguish entire industries. According to the *Mercure du Departement de la Roer*, Napoleon offered a prize of 4000 francs for the best commercial process for producing cast steel of a quality equal to that made in England. Further offers of similar premiums led to an epidemic of inventions in Essen. Krupp's brother-in-law brewed an "approved patent coffee."

Leading Continental technical experts endeavoured to solve the problem of producing cast steel, and an "Inventions Company" was even formed in Solingen. The idea was to find a mysterious chemical agent or "flux," similar to the "projection powder" of medieval alchemists. But metallurgical formulæ were less important than practical means of manufacture on a remunerative basis. Napoleon's prize was competed for in 1807 by the brothers Poncelet of Liège and by the German Fischer of Schaffhausen. Further competitors, some of whom were awarded prizes, entered later. The greatest progress in the Ruhr district was made by the "Good Hope" Ironworks. The *Westfälische Anzeiger*, in 1811, wrote: "Herr Jacoby of Starkrade, a well-known and expert ironmaster, is in possession of the secret process, which he has been using for several years past without seeking a patent on it."

Friedrich Krupp was not the German inventor of cast steel, although this distinction has been erroneously attributed to him by popular legend. Even while inventors' prizes were being awarded, its manufacture was unknown to him. In all probability he had not even made a serious effort to grapple with the problem and the first inducement to do so was the appearance on the scene of the brothers von Kechel. These retired Nassau officers claimed to possess the secret of making cast steel.

The unstable young trader, who "had no further interest in groceries," joined them in their experiments and in 1811 proceeded to build the "Fried. Krupp Steel Foundry" on the "Walkmühle" property purchased from his brother.

The optimistic founder soon came up against unexpected difficulties and dissensions. Essen chronicles blame the Kechels, describing them as incompetent adventurers. This is unfair, as Krupp learned the elements of making cast steel from them; he watched them building a fireproof crucible, designing a practicable "feed-in," and providing the furnace with proper firing.

His whole conduct pointed to reckless excess of zeal and he placed orders which remained a liability on him for years to come. In the midst of these activities the young manufacturer's eldest son was born on April 26th, 1812. The brothers von Kechel stood godfather to little Alfred, the future Cannon King.

Prolonged and expensive tests finally proved that the great expectations were not to be realized. Krupp had invested some 30,000 thalers in the venture and had equipped a complete factory with water-power, smelting furnace and crucible chamber, but success was still denied him. They succeeded only in producing cast steel in small quantities, and under the pressure of his family the disappointed Krupp decided to drop the whole proposition. The works closed down in the autumn of 1814 and the brothers von Kechel were summarily dismissed.

A year later Krupp's finances were increased by a legacy from his brother Wilhelm and by a loan from "the Jew Moses." He once more engaged a technical assistant, the retired cavalry captain Friedrich Nicolai. The latter held a royal charter giving him the sole right to make cast steel between the Elbe and the Rhine, but this "sole right" applied only to a particular method of filling the crucible and it is not clear whether Krupp took the trouble to investigate the patent properly. In any case he threw all his energies eagerly into the new venture and the works were reopened under the name of "Nicolai and Krupp." Again the partners fell out, and Krupp berated Nicolai as quack and swindler. In reality it was merely a case of an industrial speculator who, as well as his partner, lacked tenacity and sober commonsense. Nicolai actually contributed a few thousand thalers in cash and brought in a number of orders for steel, which, in view of Krupp's reduced fortune, constituted a substantial addition to their joint resources. Nevertheless he was accused of being a drone, sponging on his good-natured partner. The disputes between the two led to violent scenes, and things came to such a pass that the workpeople, whose wages were in arrears, attempted to assault Nicolai. The police were called in and Nicolai hinted at a murderous conspiracy. In July, 1816, Krupp closed down the works again.

For generations past Essen publications have declared without contradiction that the early failures of the founder of the works were due to his incompetent coworkers. Dozens of biographers repeat this statement. Some light is shed on the matter by a contemporary account of a proposal to establish a factory in Moers. As the left bank of the Rhine, since the Peace of Lunéville, belonged to France, certain opportunities for the evasion of the high import duties presented themselves. This inspired Krupp to open a file factory in Moers. He devoted all his reckless zeal to the project and in order to have the assistance of a partner with expert local knowledge he invited the young merchant Friedrich Diergardt, who had spent a short time in the Flax Market shop, to join him. Diergardt's father, rector of Moers, took a keen interest in the scheme and both father and son devoted much time and money to it. To their amazement, however, they suddenly found Krupp losing interest and their repeated letters to him remained unanswered. The changeable Essener had suddenly come to his senses and did not trouble even to reply, so the project fell through. The disgruntled partner subsequently became a prominent textile manufacturer and died as Baron von Diergardt, for which reason Krupp biographers have been far kinder to him than to other partners, who are shown as boasters and swindlers. But the accusers omit to tell us how Krupp acquired his metallurgical and manufacturing experience, as to do so they must admit that he ruthlessly exploited his earliest collaborators.

The fate of the Nicolai venture is typical of the history of German industry. The impoverishment due to the Thirty Years' War ruined handicrafts throughout the Empire and no driving force for the creation of industries was left in them. Money for financing industrial schemes was concentrated in the hands of the wealthy mercantile community.

The earliest German industrialists were rarely inventors but almost invariably financiers. In contrast to England and France the technical expert in Germany played a secondary part. If he did succeed in becoming a manufacturer, he soon succumbed to the competition of wealthier rivals. This is shown in the betrayal of Dinnendahl by the Good Hope Ironworks. Once they had sucked the brains of this brilliant engine-builder, they ruined him by deadly competition. Friedrich Krupp also displayed few scruples in ridding himself of his technical mentors.

THE BANKRUPT

In 1805 Prussia abandoned her military alliance with Austria and Russia and in exchange for Hanover ceded Cleves and Wesel to Napoleon. These territories were the nucleus of the Grand Duchy of Berg. Marshal Murat, married to beautiful Caroline Bonaparte, became, as "Joachim, Prince and High Admiral of France," the ruler of this bizarre State which was a military and political bridgehead against Germany. Murat did all he could to extend his territory, as his imperial brother-in-law had promised him an annual revenue of seven florins per head of the population. Failing in his attempts to secure more subjects, he decided to act on his own initiative. His territories were adjacent to the three Imperial Abbeys of Essen, Werden and Elten, and Murat proceeded to march troops into them under the pretext that they were formerly part of Cleves. An unexpected onslaught by Prussian troops who scaled the city walls of Essen under cover of night, threw Murat out, and "L'affaire des trois abbayes" threatened to become an important political issue. Napoleon was furious at Murat, but the latter pointed out that deputations of leading citizens of Essen and Werden had petitioned him to incorporate their cities in the French territory. Political considerations compelled the Emperor to reject the petition: "I think it ridiculous, that you should oppose me with the opinions of these Westphalians. Who cares what the peasants think, in political matters?" Murat's hopes were not realised until after the Peace of Tilsit when, in January, 1808, Essen was taken up into the Grand Duchy of Berg.

Foreign rule pressed hard on the Ruhr district, even though France, in casting off her medieval trammels, had abolished serfdom, corporations and guilds, prohibited religious persecution, reformed the legal code, decreed a progressive policy for coal mining and established the equality of all persons, thereby granting all the citizens of Essen the prefix of "Herr" (Mr.) to their names for the first time in the history of the city. All this, however, was offset by grinding taxes, a tobacco monopoly, the Continental Embargo, the continuous conscription which robbed the country of its sons and sent them to perish on all the battlefields of Europe.

There was no question of the consent of the inhabitants. Essen was governed by a mayor, two deputies and sixteen municipal councillors who were not elected by the citizens but nominated by the French prefect and deposed by him if necessary. They were merely the fig-leaf of a thinly veiled autocracy.

In 1812, the mayor of Essen had difficulty in finding a successor for an outgoing municipal councillor. There was intense unrest throughout defeated Germany, more especially in the west, with frequent riots and intensive nationalist propaganda all over the country. Under such circumstances three successive candidates had already refused the appointment as municipal councillor: Baron von Asbeck, pleading his advanced age; Attorney Tutmann, who did not wish to be hampered in his professional work; and Krupp's brother-in-law, von Müller, who stated that he was still in active service as a Prussian officer.

Friedrich Krupp was now approached. He could have given a good reason for declining, as he was under the legal age, but the young "steel manufacturer" chose to accept the appointment. He agreed to declare himself three years older than his real age and be nominated as official candidate, duly taking the oath of allegiance to the Emperor Napoleon.

The collection of Friedrich Krupp's letters and documents, two hundred and fifty pages of them, given out by the Krupp family, does *not* include a municipal decree of December 17th, 1812 (although the original is preserved in the municipal records of Essen) appointing Friedrich Krupp a city councillor and recording his allegiance to the French Emperor.

The date, December, 1812, is interesting. Inspired by the Russian disaster, a wave of national patriotism swept Germany, but it left Friedrich Krupp unmoved. Berdrow certifies that he conscientiously fulfilled his duties as municipal councillor. He had to concern himself with billeting the French troops in citizens' homes (parents of the numerous deserters being especially penalized by heavy quotas) and, adorned with the red and white cockade, Friedrich Krupp carried out the punitive measures of the foreign administration. In his "History of the Grand Duchy of Berg," Goecke states that the passive resistance of the masses was in contrast to the toadying of the officers. The nobility and men of property backed the foreign ruler with an utter disregard of their national dignity and Municipal Councillor Krupp was one of their number. This fact stands out painfully in the history of Essen, although attempts are made to gloss it over, to deny that the family had "French tendencies" and to pretend that Krupp as "a man and a German" resisted Napoleonic oppression. Furthermore, it is significant that, apart from his functions on the city council, Krupp was closely associated with the administration of the Essen Small Arms Factory and actually lent money for this undertaking to the Frenchmen Pieul and Pelletier.

In the spring of 1813 revolts of the workers broke out in Remscheid, Solingen, Elberfeld and Barmen; in Düsseldorf numerous death sentences by court martial were carried out. The days of French rule on the Rhine were numbered and the troops in Berg deserted *en masse* to the Allied opponents of France. But Municipal Councillor Krupp remained true to the tottering *régime* as long as it held power, and was one of the handful who were mobilized when all the dams were cracking. In the autumn of 1813, when Blücher's army was already approaching, Krupp helped to dig trenches for the French before Wesel.

One might speak of heroic devotion to Napoleon, if Krupp had not displayed such disconcerting haste in turning his coat; when the Allied troops occupied Essen he immediately offered his services. The municipal councillor became a city councillor and the French billeting officer turned into a Prussian Landsturm adjutant.

In the late autumn of 1816, Friedrich Krupp once more took over the direction of the steel foundry. But an evil star still hung over the enterprise, and the time was not yet ripe for an industrial awakening in Germany. The war which had devastated Europe for two decades, destroying frontiers and wiping out a generation, was indeed ended, but the expectations of those who looked for an immediate revival of economic life, trade, and industry were soon disappointed. The Continent was impoverished and immediately peace was concluded Britain swamped it with goods that were frequently sold under cost. The position of young German industry was almost hopeless, and it was further weakened by the effect of the reactionary policy oppressing the Empire now split up into more than thirty separate despotisms. The Germany of those pre-March days lacked every political or economic essential for industrial recovery.

At first the situation appeared to improve a little as the revival of handicrafts created a demand for cast steel. Krupp attempted to compete with his powerful British rivals by marking his products with the trade marks of British firms, and in 1817 the works secured a good customer in the Prussian mint. The director of the West German branch, Herr Nölle of Düsseldorf, became a friend of the young manufacturer and did his best to help him. Nölle's friendship was really the chance of a lifetime for Krupp, but he failed to see it. The only modest success achieved was in 1818 when the Gute Hoffnung Ironworks and the Dinnendahl Works both showed a substantial profit, the last-named then employing sixty workmen. This activity was largely due to the war indemnity payments flowing into Germany, which encouraged speculation.

But restless Krupp was still dissatisfied for he lacked the perseverance to build up a business step by step. Despite the improvement in turnover, he endeavoured to put up his prices, which caused protests from customers. If reality opposed his pleasant dreams—away with it, and the young manufacturer evolved schemes that promised a higher return. He would build larger works. He already owed 30,000 thalers to his family, but his mother (who inherited her son Wilhelm's money) and his brother-in-law von Müller advanced further sums, and some of the inherited land was sold. The site of the new works was to be the district beyond the Limbecker Gate where there was land which had been in the hands of the Krupp family for generations. As usual Krupp did things on a grand scale, laying out space for sixty smelting furnaces, of which only eight were actually built as this number sufficed for present requirements. The building was completed in 1819 and a further sum of 23,000 thalers was expended. The accountant Grevel had been successful in getting a laudatory article on the Krupp products printed in a Frankfort newspaper, and the sales increased slightly. The royal ordnance factories on the Rhine placed orders for steel for bayonets and gun-barrels.

Once more, in 1823, the works reached a substantial output, of which two-thirds went to the mint; but it was only a flash in the pan, as internal decay had definitely set in. Krupp's finances were crippled by expensive building schemes and his credit was dwindling; even his own relatives refused further loans. Shortage of raw materials was brought about by lack of funds and the works were forced to use scrap metal, causing delays in the fulfilment of orders and complaints as to the quality of products. There were technical difficulties also. Grevel records that one of the chief troubles was the cracking of crucibles and the running out of molten metal. The works lacked a power hammer, entailing delays in sending work out to other factories.

Krupp, disillusioned, began to lose interest and the accountant had frequently to fetch him out of a tavern to sign papers. Grevel took sole charge of production and of the business side of the works and even tried to raise money for them. But even this last helper met with shameful ingratitude, which Krupp biographers attempt to discount by disparaging the man instead of giving weight to his reproaches. The works vainly sought help from the authorities and the struggle for a subsidy runs through the history of Krupp like a scarlet thread. Applications for a grant of some 20,000 thalers were refused in 1817, 1818, 1819, and in 1823 finally. The Berlin authorities even established a steel foundry of their own and Krupp vainly offered his services to its directorate. Utterly disgusted, he made two separate offers to the Imperial Russian Government to establish a State-aided steel foundry in Russia. But these also were declined.

1824 saw the beginning of the end. The relatives were

worrying about security for their loans and his mother had to intervene on several occasions to smooth things over. In April Krupp's father-in-law Wilhelmi obtained judgment for the immediate repayment of 14,500 thalers and a settlement was arranged under which the house on Flax Market, which had been owned by the family for the past century, and may be considered as the original home of the younger Krupps, was made over to him. At a later date it was put up for sale and was purchased by the parents of the manufacturer Grillo. Krupp, with his wife and family, moved into the manager's house at the works, under the pretext that country air was better for his health and that he would thereby be brought into closer touch with the factory. He made no reference to the dire necessity which had caused this removal, neither did he make reference to the roomier and more comfortable residence on the Walkmühle estate, which was let to an acquaintance.

The Krupp family did not bear their troubles philosophically and sharp internal dissensions broke out. Krupp resigned his office as city councillor, after discharging it, not without ability, for a period of fourteen years. His name had already disappeared from the list of taxpayers, and in 1825 it was erased from the Register of Privileged Merchants, thus marking the final eclipse of the once wealthy man. The steel foundry had already swallowed up some 80,000 thalers.

Friedrich Krupp died of pleurisy on October 8th, 1826, at the early age of thirty-nine.

His name lived on in the firm "Fried. Krupp," completely overshadowed by that of his great son. He had no outstanding qualities to distinguish him from other Essen aristocrats. The one deed of his life, his establishment of the steel foundry, acquired prominence only by the growing importance of the factory half a century later. It is merely a pious family legend, that Friedrich Krupp foresaw the vast future of cast steel. He hoped for a moderately sound business, as steel was an expensive specialty for which no mass production could be anticipated. His enterprise in venturing into what was to him an entirely new field is astonishing, but he lacked all the qualifications for a sound industrialist, being neither inventor, ironmaster, nor even a really good man of business.

His day was a period of transition from the Napoleonic wars to the nineteenth century with its undreamed-of industrial progress. The steam engine had just made its appearance in industry and when Friedrich Krupp died the first railway carried coal from the Ruhr and the first steamship plied on the waters of the Rhine.

PART III

THE CANNON KING

The Coming of Steam—Other People's Ideas—The Boom in Rails—The First Gun—Blood and Iron—On Both Sides—Officer of the Legion of Honour—"The Home of His Ancestors"—The Industrial Crisis—Bloody International—Social Anxieties—Last Struggles—Lonely Death

THE CANNON KING

THE COMING OF STEAM

THE fourteen-year-old boy into whose hands the fate of the steel foundry had been committed was the antithesis of his father. Friedrich Krupp had wide vision but a weak will, while Alfred Krupp had limited imagination but a strong will.

The inheritance was certainly one of dubious value, as there were liabilities amounting to 25,000 thalers whereas the total assets of the steel foundry did not exceed 15,000. Therese Krupp took over the property in her own name, refusing to burden her young children, Ida, Alfred, Hermann, and Friedrich, with the risks attached. Only a few days after the death of her husband newspapers and letters to business associates announced that the works were to be carried on. The deceased was stated to have given "the secret process for casting steel" to his eldest son.

The business now secured what its founder failed to give it—a fixed policy. Alfred spent all his time at the works, watching production, directing and learning simultaneously. He examined raw materials, tested every possible form of crucible construction, and studied the metal being cast with the quiet perseverance which was his outstanding characteristic. All this did not make him a scientifically trained technical expert, but he did acquire a thorough grounding in foundry practice and an empirical knowledge of metallurgy. He took particular pains to ensure meticulously correct form of charge for the crucibles; and he also sought to propitiate the customers of the firm (much reduced in number and not too trustful) by conscientious service. A firm hand was at last bringing order into what remained of the family possessions.

Young Krupp soon showed exceptional commercial ability, visiting smithies and forges in the surrounding country to solicit orders. On the Enneperstrasse he came across gunsmiths who were making musket-barrels of forged steel, and, as by this time the arms industry had wholly passed from Essen to Mühlheim, this circumstance was significant. Hitherto Essen steel had only been sold as a semi-manufactured product or made into coin dies and smiths' tools. Alfred strove to raise the quality of his production, taking the most important step in that direction, the shift to rolling-mill fabrication. This doubled his returns, his yearly turnover reaching 3000 thalers, more than his father could show in his best years.

A slight improvement set in after 1830. A wave of unrest swept over Europe, the era of the Holy Alliance received its first blow. In economically backward Germany it was only in the west that disorders broke out, leading to the destruction of machines. The inclusion of the Electorate of Hesse in the North German Customs Union offered a greatly increased scope for Rhenish industries and young Krupp made a successful business trip through the valleys of the Main and the Neckar. As a result the turnover in 1831 rose from 600 to 1600 thalers. The works now employed eleven men, and Hermann Krupp, who had just completed his apprenticeship in Solingen, joined the management.

ALFRED KRUPP
1812—1887

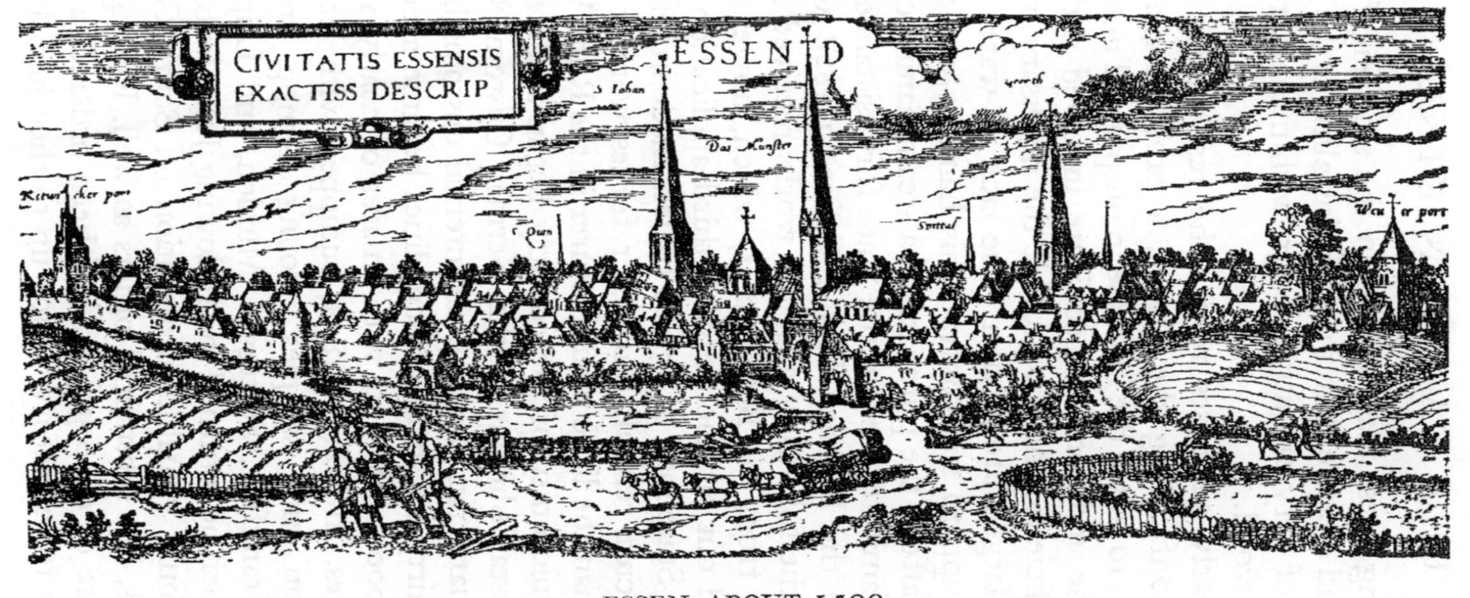

ESSEN, ABOUT 1500.
(From a drawing of the period)

A forging press was erected in the smelting furnace building and a power hammer was installed at the Walkmühle, all of very primitive design with machinery built largely of wood.

These early successes reveal the real nature of the problem confronting the works, a problem that took many years to solve; the increase in output did not suffice to meet the additional expenditure involved and there were insufficient funds to provide the necessary machinery for enlarged production. The lack of water power forced Krupp to sub-contract his hammer-work as he could not afford the purchase of a steam hammer. Once more the family turned to the State for assistance, but a petition to the Prussian Home Office in 1828 was rejected. In the following year the works received a small trial order from the Berlin mint, which, however, was not carried out satisfactorily. A third application to the King for a State loan of 15,000 thalers made in 1830 was met by a curt refusal. But other forces were at work, stronger than the little manufacturer in the Ruhr Valley; the economic unity of the country began to take shape with the establishment of the German Customs Union.

On January 1st, 1833-34, all internal customs barriers fell and thirty-six small principalities became a country of 30,000,000 inhabitants with an area of nearly 200,000 square miles. In March, 1834, equipped with letters of introduction and samples, young Krupp travelled through Frankfurt, Stuttgart, Munich, and Leipzig to Berlin. When he returned the following June, he brought back not only valuable information on South German handicrafts, but substantial orders as well. Rolling-plant to the value of 6000 gulden had been ordered, a heavy assignment for a factory which up to then had shown a turnover of not more than two to three thousand thalers. It was an order which could not be filled with the means at hand, and a summer drought had closed down the hammers. But there were far better possibilities now. Krupp's cousin von Muller joined him as a sleeping partner with a substantial investment, and Krupp bought his first steam engine from the Good Hope Ironworks. It cost 5000 thalers and had 20 horsepower. The young optimist now believed that he would be able to cope with the entire country's demand for steel.

The change to steam power marks the next stage in the development of the works. The buildings erected by their founder had hitherto proved adequate, but extensions now became necessary as production and staff increased.

The growth of the works is illustrated by the following figures:—

Year.	Steel Production.	Men Employed.
1833	9,000 lbs.	11
1834	28,000 "	30
1835	50,000 "	67

This is indeed growth. But Germany's demand was of itself not great enough for Krupp and he began to look beyond her frontiers for further sales territories. He sent out representatives and quotations to Austria, Russia, Holland, Italy, and Turkey. France offered him particularly promising business. As the importation of separate rollers was prohibited, the Essen works built up valueless "sample" frames with real rollers, which were imported into France as "machinery."

The business was, however, still far from being on a thoroughly sound basis. The new plant necessitated by contemporary industrial development had cost altogether 18,000 thalers, and had absorbed the capital put up by Krupp's cousin. In February, 1835, Therese Krupp, as nominal proprietor of the works, made the eighth application for Government assistance in the form of a loan of 10,000 thalers interest-free. But this application was refused, as was a similar request addressed to the Ruhr Shipping Administration. Ironmaster Brünninghaus, who supplied iron to the works, declined to entertain a suggestion for amalgamation. This distrust, bringing chilly refusal to all advances made by Krupp, was borne of his chronic shortage of money, only partly due to the rapid growth of the works. Alfred Krupp was evidently not equal to the task of dealing with the financial side of his business. He did not understand the art of conserving his liquid resources and building up his credit. Apart from the cost of additional buildings, huge sums were spent on prolonged business journeys. He spent a great deal of time "on the road" himself, and Hermann undertook business trips to Switzerland, France, and England. It sometimes happened that the entire management were absent for weeks on end.

OTHER PEOPLE'S IDEAS

The man upon whose shoulders rested the responsibility for this development was just twenty-five years old. He was a youth who was too busy to marry, did not mix with companions of his own age and whose restless spirit drove him from the works to the office and from his desk to his horse to visit his customers. He was, as he himself used to say in later years, "commercial traveller, accountant, smith, smelter, cokemaker, night watchman on the furnaces and everything else." His biographers are at pains to hail him as an inventor, a claim which is wholly disproved by the story of the manufacture of rollers in Essen, the chief product of the works' early days.

These rollers were used by goldsmiths and watchmakers and had to be extremely hard, with a smooth and homogeneous surface, but they must not be brittle. Krupp's years of experiments failed to produce satisfactory rollers and those which he made for coin minting at various times were not uniformly forged and cracked when in use. The first improvement in the quality of the Essen product was due to a tip given the young manufacturer by the metal grinder Rocholl of Barmen, who once worked with Alfred's father in the Gute Hoffnung Ironworks. He explained that the rollers hitherto produced had cracked due to defects in the grain of the metal through insufficient forging and that it was impossible to forge the steel properly if it was produced in the form of cylindrical boltstave; it would be better to produce it in square or octagonal section. "Never yet had so important a light dawned on Alfred" (Berdrow). He realised now what his own experiments had never shown him and promptly proceeded to apply it to his commercial production.

He was soon up against another problem. His rollers had hitherto been limited to a maximum diameter of four inches, as any increase in size weakened the core and caused cracking during hardening. Attempts to reproduce heavy British rollers had been unsuccessful. Berdrow relates how success finally came: "Alfred had often made cast steel annular rings for mechanics and discovered that they were used for making rollers with iron or steel cores. He tested such rollers himself and offered them to his friends . . ." Further progress was on similar lines. Worn-out rollers of British or Viennese make were purchased for experimental purposes from Stieber in Roth, and Krupp picked up an effective hardening process from the South German silverworkers, which he put into immediate practice at home. All these "inventions" for the manufacture of rollers were, without exception, the product of other people's brains or secrets learned and copied from the firm's own customers.

This is clearly demonstrated by the construction of the spoon roller, which was the most important achievement of these years and gave rise to a full cycle of legends. Ehrenberg writes touchingly of the years spent by young Alfred in attempting to build it, undeterred by repeated failures, in the firm conviction that he would, in due course, overcome the defects in design. The truth of the matter is that a goldsmith and engraver named Wiemer, who lived in Munich in the thirties, had, from time to time, purchased cast steel and rollers from Essen. He was an inventive designer who once built a plate-engraving machine. In the summer of 1838 Wiemer emigrated to Mexico, after providing himself with new tools and special machines. He placed an order in Essen for rollers of a quite unusual design and of exceptional size. Hermann and Fritz Krupp were keenly interested in the mysterious thing. They carefully investigated his design, discovering that Wiemer used specially engraved rollers to make spoons and forks by machinery, which until now had been produced only by hand. Alfred was absent in France at the time and replied to a communication from Hermann by saying that the construction should be carefully noted, the process might have a potential value. Alfred Krupp had no scruples about appropriating Wiemer's invention and even before he had seen the finished spoon rollers himself he offered similar ones to a Belgian firm. Meanwhile the works effected improvements in the machine and it is of importance to note that these were due to the sole efforts of Hermann and Fritz. Alfred was concerned only with its commercial exploitation.

In singular contrast to this unhesitating theft of another's secret the young manufacturer displayed the greatest anxiety about his own methods of production. Although his staff of employees was small and included experienced workmen of many years' service, he was perpetually worried about their trustworthiness. "Even the night watchman is not above suspicion; the man is about the works far too much during the day; he had better have somebody else to keep an eye on him during the night, with still another man to watch the second one!"

He was always on the lookout to safeguard his manufacturing secrets and created two confidential departments, a hardening room and a polishing shop, where the most important special processes in the construction of rollers were carried out. Only the most trusted workpeople were admitted to these departments, the doors of which were kept carefully locked. Krupp even applied to the authorities for permission to make his workpeople take an oath of secrecy in regard to his manufacturing processes. Berlin rejected this unusual application. But Krupp made his men take the forbidden oath in secret, on their honour. He exacted this oath even from the workman Borgmann whom he himself sent out to learn trade secrets concerning the hardening of steel from his competitors.

While guarding his own production so carefully, the young manufacturer conceived an audacious plan. Despite all experiments Essen crucible steel remained unsuitable for toolmaking, and Alfred Krupp believed this to be due to the raw material, not to its treatment. He proposed to try to uncover these last secrets of steel manufacture and certain other things by going to England himself. Careful preparations were made for this spy journey to the metropolis of industry. A passport was made out in the English-sounding name of "A. Crup," a passport which said nothing whatever about his position as owner of a steel foundry. According to the memoirs of the old turner Benning, Krupp worked at the smelting furnaces for some time to roughen his hands. He hoped to obtain a workman's job in a British factory. This may not correspond to all the facts, but goes to prove how well the master's schemes were known throughout the works.

Alfred Krupp began his important journey in the summer of 1838, leaving the works in the charge of his brothers, of whom Hermann was a keen business man and Fritz a brilliant engineer. They already employed a good accountant and a sound works manager; the lean days were over. Alfred Krupp first went to Paris, where his French rapidly improved, and he visited hundreds of metal workers. He reached London in October and his six months' stay in England is commented on, quite accidentally, by the German diplomat Hermann von Mumm, who relates in his book *My Experiences on Horseback*—

"During the previous winter, I met a German named Schropp, in Liverpool. We called him the 'Baron' and he was quite young, very tall and slim, looked delicate, but was good-looking and interesting. He always wore little swan-neck spurs and was quite a gentleman. I did not hesitate to introduce him to many families of my acquaintance and we saw a good deal of one another. One day, he seemed unusually solemn and begged to have a private conversation with me. He thanked me for all my kindness in introducing him to people, but this was placing him in an awkward predicament as he was travelling under an assumed name, although he carried an official Prussian passport made out in that name. His father was a steel manufacturer in Essen and as the state of this industry in Germany was behind that of England, he had come to learn the language and to try to pick up information in British steel works. His name was Krupp."

His method of obtaining such information is disclosed in a letter which he wrote to Hermann from Liverpool the end of January, 1839—

". . . as I have consistently fallen on my feet in England, which is a particular dispensation of Providence, I believe that God will further guide me to my entire satisfaction. Only yesterday I was walking with Fritz Sölling five miles from here and went all over a recently opened rolling mill for copper plates, without any introduction, although nobody is allowed to enter. I was well booted and spurred and the proprietor was flattered that two such smart friends should deign to visit his works."

J. A. Henckel, the founder of the Solingen-Twin Works, wrote from Sheffield at about the same time, in a very different strain—

"We visited a steam grinding mill and several worked by water power and, being business men, we hope to be able to visit any works which we may wish to see. Our inspections continue from ten o'clock in the morning until ten at night."

Henckel visited factories, including Sheffield makers of his own range of products, quite openly. He filled his diary with notes on what he observed. Alfred Krupp avoided the straight way and preferred a devious course.

The young manufacturer's visit to England taught him many things and he is frequently quoted as having said that it was his stay in England which first opened his eyes to the immense sales a really good article might enjoy. He told Hermann he was now sure that the quality of steel depended on the iron from which it was made. He returned home through Paris and Belgium.

Meanwhile business conditions began to go down as the crisis of 1839 engulfed many customers in Essen, the smaller independent artisans being forced out of business by industrialists. Krupp returned to the works after an absence of fifteen months to find all kinds of trouble awaiting him. An able traveller of his, who had established good contacts in Berlin and Moscow, had died. Orders were scarce and money hard to come by, while the Herstatt Bank in Cologne was pressing him. The family still had reserves of capital and 3000 thalers were realized by the sale of two of their properties.

Alfred Krupp remained in Essen a few weeks only. He gave instructions in the use of the Swedish ores he had brought from England, then went off again on journeys that stretched over several years. An odd haste lay over his business dealings. He hurried from Berlin to

Vienna, from Vienna to Warsaw, without returning to Essen betweenwhiles. He was no longer concerned with single rollers but with complete rolling mills. One such plant, worth over 5000 thalers, was ordered by the firm of Vollgold of Berlin in 1840.

Krupp's lack of technical vision is clearly illustrated by an incident which occurred at about that time and which brought his business to the verge of bankruptcy. After a lengthy period of business contacts the Vienna mint placed a huge order with the Essen firm: the construction of a sizing-works with five machines, thirty-two rollers and other new adjustments. It was to cost nearly 30,000 gulden, a high stake for conditions in Essen. But Vienna had written stringent conditions into the contract. Especially important were the specifications regarding sizing of different denominations of coins, something that had hitherto "not been accomplished by any other mint."

Alfred Krupp signed this contract without noticing, apparently, that it imposed impossible demands on the precision of the rollers. He promised what he could not carry out. The Mint authorities took his promises seriously. But when delivery was made their engineers found that the specifications had not been complied with and insisted on a rolling mill the product of which did not need later sizing. Krupp tried to improve on his completed work, but was soon compelled to admit that he was unable to meet the requirements of the mint authorities. Whereupon the Mint refused payment.

Krupp maintained that their attitude was due to malice and prejudice inspired by his competitors. He talked of sabotage and stated that his rollers had been damaged by excessive pressure—an unfounded accusation. The dispute lasted a whole year and non-payment of the money threatened to ruin the business. Finally, von Kübek, president of the exchequer department, took pity on the young manufacturer and arranged for a second delivery, free from defects, to be paid for in part immediately, the balance being paid subject to the Emperor's approval.

A fresh order from Vienna brought little comfort, and the complaints made about the way it was carried out cause Krupp biographers to comment scathingly on "the sharp practice and bad faith of the customer, whom Krupp described as a trickster of the worst type."

In the desperation of the Vienna conflict Krupp began to toy with an idea of which we shall hear much from now on. He had become acquainted with an agent of the Russian Ministry of Finance dealing with industrial matters in Russia. This man proposed the opening of a branch factory in Russia. Krupp wrote from Vienna—

"The Prussian Government have done nothing for me, so I cannot be considered ungrateful if I decide to leave my own country for another, whose authorities possess the wisdom to foster industry in every possible way." Emigration to Russia had been a pet idea of Friedrich Krupp's. And his son was equally ready, at any serious disappointment, to shake the dust of his own country from his feet.

The spoon-rolling mill, however, led to important and successful business. In 1843 Krupp associated himself with the Vienna branch of the old Rhenish manufacturing family of Schöller. Works for the manufacture of table cutlery and plated sheets were built at Berndorf, near Vienna, with Schöller in charge of the commercial and Krupp directing the technical side of the business. The latter acquired a holding equal to

Schöller's only on investing the same amount of capital, which he did in the next few years. But here also disputes arose between the partners. Alfred Krupp gave a verbal assurance that he would refrain from selling spoon-making plant elsewhere until the Berndorf factory was on its feet. When, therefore, he discussed schemes for further cutlery works in Berlin and in the Rhenish Provinces, Schöller not unnaturally objected. Krupp retorted that this promise was intended to protect the Essen and not the Berndorf Works. Although Krupp was obviously in the wrong the conflict died down as the German schemes failed to materialize.

The prospective partner for these schemes was the Elberfeld manufacturer Jäger, who was closely related to the local bankers von der Heydt. The connection would have been of great value to a business that suffered perpetual financial distress. But Krupp's distrustful egotism rendered him incapable of loyal co-operation. He invariably expected others to resort to the underhand methods and double dealing which came naturally to him. He learned that the products of Jäger's Works included cuirasses (breastplates for cuirassiers), which were not advertised as being made from Essen steel and he immediately proceeded to threaten and insult the man whom he believed to be prejudicing his interests.

THE BOOM IN RAILS

Early in the forties the works increased their range of products to include the complete equipment of rolling mills, machinery parts and steel springs. The four workmen of 1826 had now grown to ninety-nine and, although the old smelting furnace building still formed the centre of the works, it was surrounded by a whole series of new buildings, boiler house, grinding shop, smithy, and storehouse. A new dwelling-house was being built alongside the old residence which the family had occupied for the past eighteen years.

The one thing lacking was quiet and consistent development. Towards the close of 1843 a serious lack of funds was again experienced, and the resources of Krupp's cousin were so depleted as to force him to mortgage his entire property with a Cologne bank. Alfred's boyhood friend, Fritz Solling, who owned a large mercantile business, came to his help with loans and eventually became a silent partner with an investment of 50,000 thalers when Müller had dropped out. Sölling was a member of an Essen family which had been related to the Krupps centuries earlier. He was a shrewd accountant, knew the foreign markets, and became an indefatigable critic of the Krupp business methods. He disapproved of expensive journeys, continuous new building, and the squandering of working capital. There was some serious friction when, in accordance with circumstances provided for in the partnership agreement, he was called upon to increase his investment to 75,000 thalers, and Krupp was compelled to write a number of letters to his Cologne partner before the latter reluctantly put up this additional money.

The high protective duties set up to assist German industry led to increased activities in Essen. Orders from Berndorf and business in mining equipment made the tall chimneys smoke. The number of Krupp's employees rose to 109 in 1844 and to 124 in the following year. As works manager he put in his cousin Ascherfeld, a man of primitive mind who believed in the rule of rod. Iron discipline became the order of the day, and when the factory hands were seen hurrying to work in the early morning, passers-by called mockingly: "Better hurry, bell's ringing!"

The foundry had now gained wide experience in the casting of steel and was familiar with a whole range of different grades. Nevertheless Krupp was by no means in the front rank of contemporary research workers in this field. The French metallurgist Le Play had already published the first scientific treatise on steel foundry practice.

One event of this period opened up a new sphere of activity for Krupp—the Berlin Exhibition of Industries, where he embarked on an intense publicity campaign. The German Customs Union displayed samples of manufactured goods from all its territories and Essen had a wide assortment, such as rolling plant for gold and tinsel, a chime of tubular bells designed by Fritz Krupp, and hollow forged cold-drawn musket-barrels.

Alfred Krupp paid another lengthy visit to England, where he secured a patent on his spoon-roller, and to France, where he had an introduction to James von Rothschild. Serious news brought him home again. A great industrial slump had set in and the 1846 turnover of 80,000 thalers shrank to half that amount in the following year. The balance sheet showed a heavy loss and Sölling declined help as he had long since prophesied some such trouble. After a struggle of twenty-two years the works trembled to their foundation.

The Krupp family made a lamentable showing in this hour of danger; they started quarrelling amongst themselves. There never had been any great unity among them and their letters of that time reveal rivalry between the brothers. Alfred was at pains to retain the sole direction of the works, while Hermann appears to have been a really able business man with a less gloomy disposition than his elder brother. His letters, containing valuable advice, indicate that he was a manufacturer of considerable experience. The youngest brother, Fritz, was undoubtedly the most technically able of the three. He had raised production by a number of ingenious improvements and had carried out several important inventive experiments. He designed the tubular chimes exhibited in Berlin and made the first experiment with steel springs. In his leisure hours he was engaged on the construction of a vacuum cleaner and of a self-propelled carriage. It is difficult to determine whether his experiments were prejudicial to the business, as Alfred claimed.

The latter, self-centered, utterly lacking in consideration for others, cannot be regarded as a fair judge of his brother.

Both his mother and Fritz Sölling were on Alfred's side. They thought that his age and greater energy would make for a safer policy in the growth of the works. As Therese Krupp-Wilhelmi was still the nominal proprietress of the business she tried to get the brothers and sisters to agree to a transfer of the ownership to Alfred. Ida and Fritz were to receive compensation and Hermann was to take over the partnership in the Berndorf Works. Fritz raised objections. He knew that his brother was really seeking to turn him entirely out of their father's business. The mother then proceeded to make her will, under which she made Fritz's financial holding in the works conditional on his not disclosing any of their business secrets; nor founding or assisting in the management of any similar business which might become a competitor. The aggrieved and indignant Fritz left Essen and made various unsuccessful attempts to start new works, finally settling down in Bonn.

On February 24th, 1848 (the day the Paris revolution broke out), Alfred Krupp entered into sole possession of the works for the amazingly low price of 40,000 thalers, taking over all assets and liabilities. And, as the Berndorf Works reported a loss, which, according to their agreement, had to be shared by Essen, the consideration was actually reduced to 25,000 thalers. The young man was now master in his own house, even if the house did not rest on very secure foundations. But, Alfred Krupp had what is indispensable to any successful venture, namely, luck! From the moment he became independent, the affairs of the business began to prosper.

It was a time of general unrest and labour troubles throughout Europe, but Alfred Krupp's horizon was bounded by the interests of his foundry. He looked upon industrial unrest of any kind merely as a hindrance to business. Ida Krupp writes in a letter of those days: "Alfred assembled the workpeople yesterday and spoke of the general unrest, which he hoped would not spread to Essen, but should it do so, he expected his men to do all they could to stem it." To preserve his working force from revolutionary contagion, Krupp kept them all day long in the foundry. If no orders came in, they would be set to cleaning up. Then in the late evening the truculent reactionary Ascherfeld marched the men back to the city, the gates of which, closed through the proclamation of a state of siege, were specially opened to admit them.

Despite this isolation, there were murmurings amongst

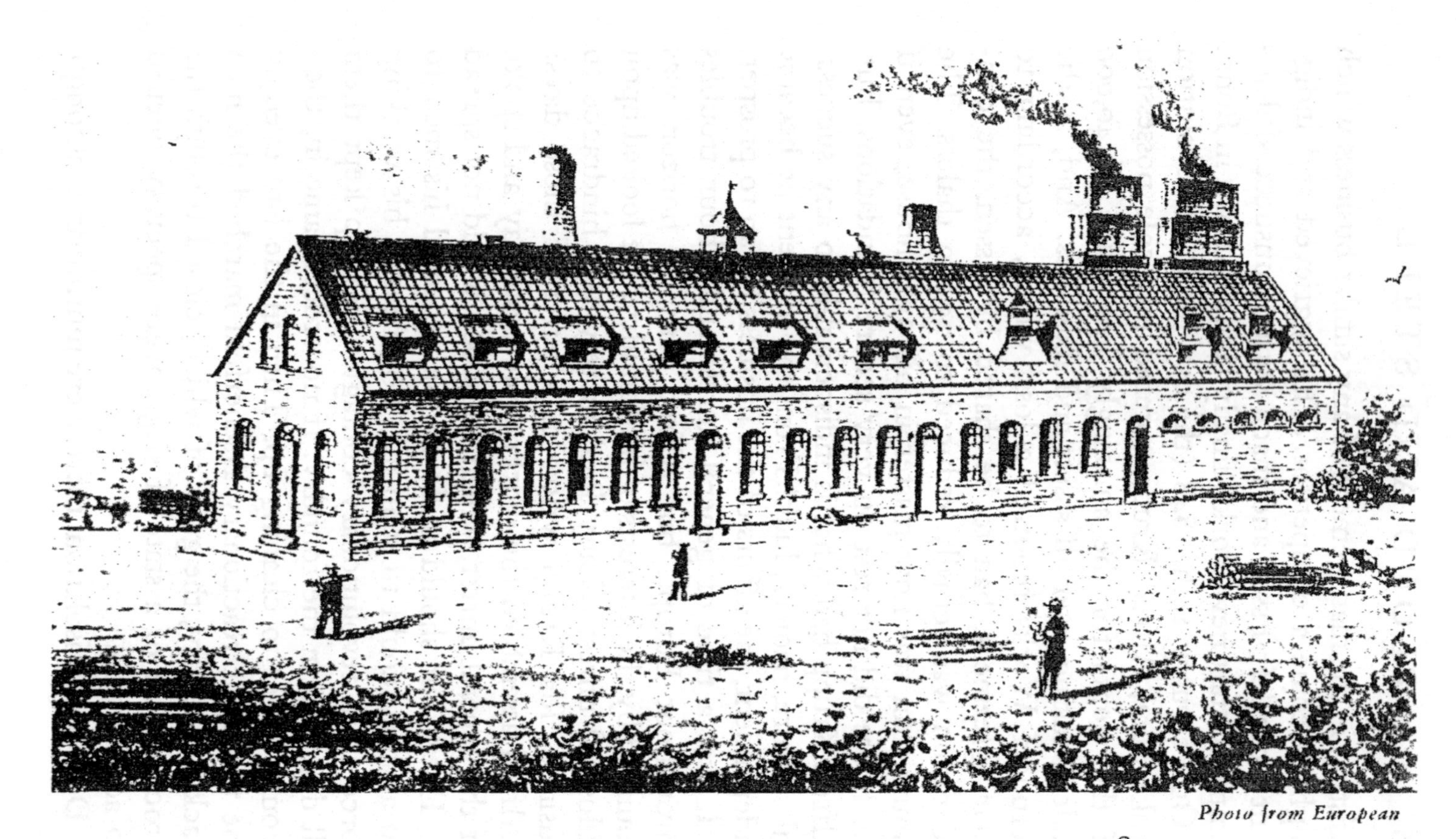

Photo from European

THE FOUNDRY ESTABLISHED BY FRIEDRICH KRUPP IN 1811.

THE SO-CALLED "ANCESTRAL HOME."
(Rebuilt in 1872)

the employees. The smith Marré, the last of the old guard whom Alfred found in the works when his father died, dared to make complaints. Krupp looked upon this as insubordination and summarily dismissed the man. Similar treatment was meted out to the old fitter Huelsmann. With undisguised brutality Krupp showed that he was master in his own house. One of the many Krupp legends pretends that, notwithstanding a lack of orders in the year of the revolution, there were no dismissals, whereas in actual fact the number of employees was reduced from 122 to 74, the services of highly trained key-men being, of course, retained whatever the state of the business.

The financial crisis of that year led to numerous bank failures and the Cologne "Schaffhausen Bank" closed its doors, Krupp being driven to raise money by the sale of the family plate. Sölling's guarantee enabled him to open a credit with the Cologne banking firm of Salomon Oppenheim, but repeated overdrafts led to disputes which served to provide Krupp with a lifelong grievance against all banks. However, money began to come in again when the Duke of Leuchtenberg, grandson of the Empress Josephine, bought a spoon-rolling plant for over 20,000 roubles and a second was purchased by the Birmingham firm of Elkington, Mason & Co., who paid £8000 for it together with the sole selling rights in the United Kingdom. These transactions eased the financial pressure on Krupp. The spoon-rolling mill in its improved form had been the means of consolidating his position and the fact that it was the creation of his brother justified Fritz Krupp's declaration that he had been swindled.

The factory now secured breathing space for further

research work. Business with rollers fell off, as the industrial importance of railways became more apparent. The first German railways were built about the middle of the century, the Cologne-Minden line being opened for traffic in 1847. Essen's railway station was built some years later. The mileage of the Prussian railways increased twentyfold in the following thirty years and a new field was opened to industry. In common with other manufacturers Krupp saw that railway orders constituted a turning point in the fortunes of his works.

German industry lived through a period of growing prosperity from 1850 to 1856, following the opening of the California gold fields. The output of the Essen Works rose steadily until their annual orders for spring buffers and railway carriage axles reached many thousands. A single order was now frequently worth 100,000 thalers. The works were extended to eight times their original size and now included a fitting shop, the long-desired rolling mill, power-press, puddling furnace, and iron foundry, and large power hammers thudded noisily. The staff organization was altered also. Gantersweiler, the travelling representative, was made general manager and Ascherfeld technical manager. Satisfactory arrangements with the banking houses of Mendelssohn & Co. and Bleichroeder ensured financial stability, and silent partnerships, which brought in 250,000 thalers, were granted to Niemann in Horst and the brothers Waldthausen.

Alfred Krupp was not content to await railway orders, but cast about for ways of securing further ones. Acting on the advice of Director W. Lueg of the Good Hope Ironworks, he turned his attention to the tricky job of making railway car tires. The rapid increase of speed and load to which rolling stock was subjected, rendered it more and more difficult to maintain the practice of welding tires to wheel rims, causing dangerous breakage. Krupp's experience in the manufacture of rollers had taught him how to produce seamless annular rings of uniform toughness, although he first got the idea, as already related, from orders placed with him by engineers. He now proceeded to apply this knowledge by splitting cast-steel blocks and rolling them into wheel rims. He protected the process by a patent and maintained great secrecy over it. The workpeople nicknamed the screened-off portion of the shop devoted to this particular work "Siberia." Suitable high-grade material was utilized for the tires, which soon became one of the most important items of the Essen Works' total output, attaining a volume of 32,000 per annum. Ships' fittings and conveyors, orders for which were secured through Sölling's valuable connections, also proved remunerative.

Notwithstanding all these successes and the increasing recognition accorded to him, Krupp was regarded as an outsider by his fellow-industrialists. While in Berlin he got to know young Gruson, who, as shop foreman of the Hamburg railway, ordered axles for test purposes from Essen. One of these was sent to Krupp's competitors, the Karls Works in Eberswalde, for examination, and reports on the relative merits of cast steel axles from Essen and forged axles were unfavourable to Krupp, who immediately and angrily issued a warning pamphlet regarding hardened forgings. He suggested further comparative tests at Messrs. Borsig's Works in Berlin. He fell out with Gruson, who was not disposed to put up with Krupp's high-handed methods and the quarrel reacted unfavourably on the Essen Works, as Gruson was a brilliant engineer who was in a position to injure their interests. Krupp's overbearing manner irritated the numerous technical experts attending the tests at the Borsig Works and was primarily responsible for his isolation, for he was friendless amongst other German industrialists.

THE FIRST GUN

Krupp noted that his brother in Berndorf was advertising swords made in the Vienna branch of his factory during the Revolution year, and was annoyed to think that his own attempts to secure arms contracts had so far proved fruitless.

In spite of a long family history of connection with the trade in weapons, . . . a history of which Alfred Krupp apparently knew little, the inducement to enter the armament industry came to him, as was usual with him, from an outside influence. In 1836 Hermann Krupp wrote from Munich saying that a local gunsmith required two cast steel musket-barrels, as iron ones tended to roughen. Early in the forties Alfred proceeded to develop the idea and personally forged the barrels, although there is evidence that he induced a fitter of his acquaintance to improve on his work. He dispatched "the first mild steel musket-barrel produced" to Lieutenant von Donat of the Mülheim Small Arms Works, in 1843, and in the following March sent two more to the Prussian Minister of War, von Boyen. At the same time he offered to make a mild steel field-gun barrel.

But it is no easy matter to obtain a hearing from a Prussian bureaucrat. Not until it became known that the Government of Louis Philippe had, through the intermediary of James von Rothschild, carried out tests of Krupp's musket-barrels, did the Berlin authorities follow suit. There appeared to be ample scope for improving the barrels of small arms, as the newly adopted needle gun made demands on the material which iron barrels could not possibly satisfy. Nevertheless the Prussian War Minister sent the following typically red tape reply to Krupp, on March 23rd:—

"With reference to the offer in your letter of the 1st instant, I regret to inform you that no use can be made of it for the purpose of manufacturing firearms, as the present method of producing barrels is less costly and is so satisfactory in other respects as to render the consideration of any alternative method of manufacture quite unnecessary."

This rebuff was frequently quoted by Krupp himself as a typical example of the lack of vision on the part of the Prussian bureaucracy, a lack which was to render him great service in future years.

The above case is, however, completely eclipsed by other shining examples of official misjudgments, some of which are positively tragic; for instance, the needle gun was rejected by military experts on the plea that the increased rate of fire would waste ammunition and that firing in the lying position was prejudicial to morale. And in 1908 the Wright brothers received the following snub from the British Admiralty:—

"With reference to your communication concerning the use of aeroplanes, I have consulted my technical advisers and regret to inform you that the Admiralty are of opinion that they cannot be of any value for naval purposes."

Barely six years later German aircraft were bombing London. Krupp had less claim to pose as an unrecognized inventor, as there were serious technical objections to his barrels although he clung to his belief in hollow forged ones in the face of the generally accepted superiority of bored ones.

The case of cast steel cannon was more favourable as great changes in the construction of guns were taking place. For centuries past guns had been cast from bronze or iron. The cheaper cast steel was brittle and its use led to fatal accidents due to bursts (as was demonstrated at the siege of Sevastopol) whereas bronze was expensive and too heavy for field artillery, the importance of which was increasing year by year. The shift however was actually brought about by alterations in cannon construction. Improvements effected in small arms had increased their range and velocity to such an extent as to put hostile batteries quickly out of action, as the field guns were still muzzle loaders firing solid spherical shot with excessive clearance in the barrels and consequent inaccuracy of aim. Gunnery experts endeavoured to remedy these defects by improvements which included the covering of the round shot with lead to reduce the barrel clearance and the provision of lugs on the shot to engage in spiral grooves cut inside the barrel, the grooves being developed into a "rifling" which imparted a rotary motion to the projectile.

These improvements in construction imposed a correspondingly greater strain on the material from which the gun barrels were made and it was soon found that only cast steel was capable of standing the strain. Essen chroniclers endeavour to give Alfred Krupp the credit for being the first to advocate its use in the construction of cannon, but there is overwhelming evidence to show that the use of cast steel guns was urged by Decker in 1816 and by others after him. Nevertheless some credit is due to Krupp for his campaign against the reactionary views of the ordnance board.

In April, 1844, the Berlin military authorities ordered an experimental cast steel 3-pound gun in Essen, but the specification laid down a number of stringent conditions with which Krupp was barely able to comply. His lack of a furnace and of a power hammer big enough to deal with the 2000 lbs. casting led to delays in the construction of the gun, which was only completed in 1847, thereby losing the Essen Works their priority rights in respect of the patent as other makers of ordance entered the field against him and anticipated his delivery date.

The completed 3-pound gun had an inner tube of cast steel inside an outer one of cast iron and it remained at Spandau Arsenal for many months before being used for trials in June, 1849, under the scrutiny of a committee of artillery officers. After firing about 100 rounds the gun was deliberately burst in order to test the strength of its material. The committee expressed doubts concerning the possibility of producing heavy ordnance of uniform quality constructed on this principle and criticized the high cost involved, whereupon the War Ministry informed Krupp that unless these could be reduced, further tests would scarcely be feasible.

The cost of production was the crux of the whole matter. All Europe was in a turmoil of domestic unrest in these mid-century years; unrest which distracted the attention of all countries from warlike preparations. The age of competitive armaments at any price was still to come.

Another opportunity for Krupp to call public attention to his cast steel gun barrels was presented by the London Exhibition of 1851. The Essen Works were not slow to seize it and displayed a 6-pdr. gun of cast steel beside the pavilion over which flew the flag of Prussia, together with burnished cuirasses, rolling plant, railway carriage axles, springs, and coin dies. The main exhibit, in the centre of the pavilion, was a gigantic steel casting of 2½ tons which had to be cast from 98 crucibles and which was nearly a full ton heavier than the largest British "monsterpiece." It was an impressive exhibit for a small manufacturer. For the first time Krupp succeeded in attracting the attention of the leading European industrialists to his products. Queen Victoria and the Master General of the Ordnance stopped to admire his steel casting, and the Essen Works were duly honoured by the award of the Council's bronze medal for exhibits in the steel section. Attempts to magnify this success by statements that Krupp had overshadowed all his British fellow-exhibitors are disproved by a letter written by Prince William of Prussia, the future king and patron of Krupp, in which he says: "The British industry is possibly behind the French in regard to the exhibits of bronze; in all other respects it undoubtedly has the lead."

After smaller industrial exhibitions in Düsseldorf and Munich, the year 1855 offered Krupp another opportunity for publicity. Emperor Napoleon organized a World's Fair which was to be the event of the decade. Krupp decided to exhibit a steel casting of even vaster dimensions and a block weighing five tons was dispatched to Paris. Ingenious forwarding agents provided him with successful publicity in the shape of breakage of specially re-inforced trucks, leaving the huge steel casting stranded in the streets exposed to the admiring gaze of the public, while rollers and lifting gear were rigged for its further transport. When finally erected in the Krupp pavilion, it broke through the floor—"La sacrée tête carrée d'Allemand" was a sensation even before the exhibition opened. Krupp now surpassed himself and offered to produce steel castings 12½ tons in weight. He exhibited a cast steel gun identical in design with the new French field gun, "canon de l'empereur," but which was nearly 200 lbs. lighter than the latter, thereby immediately arousing the interest of French military circles and of the emperor. Napoleon created Alfred Krupp *Chevalier de la Legion d'Honneur*, a distinction gratefully accepted. The exhibition jury awarded him the gold medal.

Nevertheless Paris did not turn out to be an unqualified triumph for Krupp. His supremacy was challenged by a competitor who had already given him considerable trouble in the past, the Bochum Steel Foundry. Since the establishment by the Swabian Jakob Mayer of the small foundry bearing the title "Bochumer Union," these works had, under the direction of Louis Baare, developed into a formidable competitor. There had already been conflicts, more especially in regard to the Krupp patent wheel rim which was opposed by Bochum. The latter put forward the justifiable contention that Essen was merely exploiting an old process known to every smith, which had been used by Bochum for many years past. The Patent Office decided in Krupp's favour after the latter had succeeded in persuading them that Bochum had learned the Krupp process through dismissed Krupp workers. The conflict became sharper when Jakob

Mayer discovered a valuable new method of producing steel castings of any desired shape, now known as "shape casting"; thereby defeating Krupp in his own specialized branch of industry. The latter, still ignorant of metallurgy, attempted to prevent Bochum from using the designation "cast steel" for their shape cast products.

The Bochum works created a sensation at the Paris Exhibition by showing church bells cast in steel by their new process. Krupp did not hesitate to continue his attacks in the French capital. He instructed his local representative, Haas, to protest to the jury regarding the use of the designation "cast steel" by the Bochum works. He even tried to have one of the bells broken up for examination by French experts and offered to defray its cost (2100 francs).

All Krupp's manœuvres proved unavailing. The judges recognized the value of the Bochum invention and awarded it a gold medal. Krupp still continued his opposition and fell foul of the Prussian exhibitors' committee, who considered him to be both querulous and unfair. Solling wrote a scathing denunciation of him.

The Krupp works had now become a really large enterprise. The seventy men employed there when Alfred Krupp took control in 1848 had now increased to 704. There were resident agents in all the great cities of Europe: Karl Meyer in Berlin, Matthias von Ficzek (who also represented the Rothschilds) in Vienna, Henry Haas in Paris, and the able Alfred Longsdon in London. Theodor Topp now shared the general managership with Gantersweiler. King Frederick William IV bestowed the Order of the Red Eagle on Krupp. In 1851 Prussian Minister of Commerce von der Heydt visited the works and two years later came Crown Prince William.

The manufacturer now began to feel lonely. He had parted from his brothers and sister and in 1850 he lost his mother, Therese Krupp-Wilhelmi, to whom he owed a great deal although his attitude towards her had always been rather casual. She spent her last years in her small house on the outskirts of Essen, seldom visited by her son. In the spring of 1853 Alfred Krupp's friends were surprised by news of his engagement. Shortly after this, came the announcement of the wedding of the forty-year-old manufacturer to a girl half his age, daughter of the Inspector of Taxes, Eichhoff, in Cologne.

Their married life began none too well. The damp climate of Essen did not suit the health of the young wife and she was compelled to absent herself more and more frequently as time went on. Her husband's letters began by references to prevailing fashions, clothes, furs, and the wife's health, but after a few preliminary sentences they concerned themselves mainly with the works and matters of business. The manufacturer had little time for purely personal matters. In 1854 an heir was born, who was named Fritz.

Alfred Krupp was now outgrowing Essen. The passing of the years and failing health due to overwork forced him to seek rest and relaxation in health resorts. In the summer of 1855 he spent a few weeks in Pyrmont, where he was suddenly assailed by the fear of death. He sent for his cousin Ascherfeld and compelled him to enter into a secret agreement to take over the direction of the works in the event of his own death and to make written records of all the research work they had carried out together.

BLOOD AND IRON

The hour of the cast steel gun came with the wave of wars which broke out in the middle of the nineteenth century. The wars were caused by the desire for self-determination on the part of the newly awakened nations. What the year 1848 began was completed by the revolution from above; the "Europe of Treaties" created by the Congress of Vienna fell to pieces. In Prussia the year 1862 saw King William at loggerheads with his Parliament because of costly and reactionary military reforms, which resulted in the appointment of Bismarck as Prime Minister. On September 29th, in the course of a speech to the budget commission, he declared: "The solution of the great problems of these days is not to be found in speeches and majority rulings—that was the mistake in 1848-49—but in blood and iron!"

These words were symbolical of the epoch. Wars recurred year after year: in 1868 the Polish rising against Russia; in 1864 the Austro-Prussian campaign against Denmark and the Civil War in the U.S.A.; in 1865 the French intervention in Mexico; in 1866 the Italo-Prussian war with Austria; in 1870-71 the Franco-German war; in 1875 war in the Balkans; in 1877 the Russo-Turkish war. Finally in 1878 the Congress of Berlin met, and settled the last of the debatable frontiers. Those twenty years of warfare ushered in the first era of competitive armaments in the modern sense, and the time had now become ripe for a high-grade and costly product—such as the cast steel gun—to come into its own as a symbol of this age of blood and iron.

Krupp had now struggled for ten years to introduce

his gun to the unreceptive military authorities and to convince them of the superior merits of cast steel. One of his early admirers, Colonel Orges, published a brilliant appreciation which led to an order from the Prussian Government. The Harvey ordnance works in England ordered a steel casting of 6000 lbs., as Essen now had the reputation of being the only works capable of producing such blocks. In Paris a 12-pdr. gun shown at an exhibition was purchased by the military authorities for experimental work. Trial orders came from Russia and the first really large one, for thirty-six guns, from Egypt. But even these achievements were insufficient to keep the works going, as the manufacture of firearms called for costly and complicated plant which had to be continually used if it was to pay for itself. Towards the end of the fifties it became increasingly apparent that hope of support on the part of army bureaucrats was a disastrously long-term draft.

Krupp was, however, not merely a manufacturer, like his rival in Bochum, but essentially a business man. He had shrewdly appraised the difficulties of doing business with governments otherwise than by influence and nepotism. Accordingly he set his course towards the feudal-monarchistic clique which would come into power in Prussia when Crown Prince William took over the Regency. Through the good offices of the Master of the Horse Krausnick, with whom he had come in contact because of their common interest in horses, Krupp secured the friendship of Prince Charles Anthony of Hohenzollern, Divisional Commander in Düsseldorf, later appointed President of the Council of Ministers in Berlin. He became even more friendly with General Bernhard von Voights-Rhetz, A.D.C. to Prince William, later to become Director of the War Department. Both soldiers gave their unqualified support to the manufacturer who revealed himself as a staunch friend of the army, and who was, moreover, such an excellent host. On the eve of the *coup d'état* in Prussia, outwardly signalized by the advent to power of Bismarck, the Ruhr ordnance maker had powerful friends.

Their influence is demonstrated by the events of 1859. Krupp biographers pretend that the change in the attitude of the Berlin authorities towards the Essen works came as a sudden surprise. But in actual fact the rearmament of Prussia was preceded by a series of backstair intrigues. Krupp's friends informed him that the Prince Regent proposed to counteract the growing national dissatisfaction by an increase in the country's artillery. Voights-Rhetz and Prince Charles Anthony did their utmost to further their friend's interests by calling the Prince Regent's attention to his wares and received assurances every consideration should be given them. William was as good as his word, and on May 10th the Prussian Cabinet authorized the placing of the first large order with Krupp. It provided originally for seventy-two solid drawn 6-pdr. guns, but when submitted for the Prince Regent's approval the latter personally increased this figure to 300.

Krupp was master of the art of winning the confidence of the simple-minded William. He let it be understood that his firm had turned down important foreign orders from patriotic motives and that, above all, he had declined to supply guns to France. William came to regard him not merely as a business man but as a misunderstood patriot to whom the country owed a debt of gratitude, and he soon found occasion to demonstrate these sentiments publicly. In September, 1861, the great power-hammer "Fritz" was started up for the first time at Krupp's works. Its weight was 6000 lbs. and it caused a local Ruhr ironmaster (said to be Haniel) to exclaim "Has Herr Krupp gone mad!" Shortly afterwards the Prince Regent, accompanied by his son and by the Minister of War, visited the works to see "the biggest hammer in the world." Our manufacturer could now rest happy; the ministries and the ordnance department would not be likely to overlook the interest shown by the Regent in the Essen works.

William had a good reason for winning the friendship of the arms manufacturer. Shortly after his coronation he found his Parliament opposed to his reactionary plans for army reform. The people of Prussia were displaying increasingly liberal tendencies, rejecting conservative influences, and calling for a curtailment of the royal privileges. In March, 1862, the Prussian Chamber decided that in future revenue and expenditure were to be itemized in detail as the Government had used funds, earmarked for other purposes, for the Army. The King dissolved his audacious Parliament, but fresh elections increased the strength of the Progressive Party by a third of the total number of seats. After a prolonged struggle Bismarck again secured a dissolution in the spring of 1863, but this time the Progressives secured 40 per cent. of the votes and the position became hopeless as Parliament flatly refused to vote the necessary credits for the army. Bismarck rose to the occasion by finding a loophole in the Constitution, which enabled him to continue his rule without the voting of credits.

The Progressive Party had the support of all enlightened citizens, especially in the liberal-minded western Provinces, the Rhineland and Westphalia. In Essen there was a dramatic conflict between the state and municipal authorities; the Liberal Party leader Hammacher was elected to Parliament three times running and rejected each time by Berlin. Essen's struggle against the unconstitutional acts of the Government became more intense and the municipal authorities declined to celebrate the Fifty Year Jubilee of their union with Prussia. The Burgomaster and Council addressed a petition to the King begging him to dismiss the Bismarck Ministry. The petition, accusing Bismarck of preparations for war, was signed by all the leading citizens of Essen, including all its wealthy manufacturers.

Alfred Krupp alone was not among the signatories! He had decided to espouse the cause of the coming man, whom he sensed as important in future rearmament. As soon as he heard of Bismarck's trouble, he wrote to the Ministry of War offering, in view of the obstructive attitude adopted by Parliament, to give long-term credit up to two million thalers for orders placed with him. He wrote in a similar strain to War Minister von Roon on February 22nd, 1864.

Krupp was no admirer of democratic control. His ideal was despotic Russia, where an expenditure of millions could be authorized by a stroke of a pen. He saw great times coming for Prussia when Parliamentary disapproval and similar democratic nuisances were no longer allowed to exist. He was all the more eager to support this unconstitutional *régime* since it helped to ensure good business.

The King was greatly touched by Krupp's offer and would no doubt have made use of it but for the outbreak of war with Denmark. Krupp, however, was not altogether satisfied with the results of his action, as War Minister von Roon coldly declined the offer. The old Prussian soldier was a disinterested fanatic, who regarded the "Rhenish outsider with undisguised suspicion" as Berdrow writes. Years of business dealings failed to break through the inflexible reserve of the incorruptible Minister.

Further opposition was encountered from Minister of Commerce von der Heydt. He was a member of the old Elberfeld banking family of that name, which Krupp had offended in the forties in the course of his dealings with Jäger. There had been friction on various occasions and although the Minister could not actually be accused of hostility to the over-zealous manufacturer, his attitude towards him was reserved. An open conflict arose over Krupp's application for an extension of the patent on the wheel tire. The Essen cast steel rims were now coming into general use and the harvest had yet to be gleaned, but von der Heydt rejected the application for the extention. The infuriated Krupp openly accused the Minister of deliberately seeking to injure his works in order to further the interests of others with whom he was associated.

Krupp now felt strong enough to do battle with the Minister and petitioned the King, through the Minister-President direct, by the aid of Voights-Rhetz, for an extension to the life of the patent, with special reference to its importance in view of financial liabilities incurred by the Essen works through their refusal of offers to exploit the patent in foreign countries or to establish similar works abroad. Furthermore, he emphasized the financial losses which his refusal to sell cast steel guns to foreign countries had entailed for his business. Every word of this statement was a lie, but it proved effective. Voights-Rhetz mobilized Prince Charles Anthony and Minister Delbrück to support his protégé's cause before the Regent, with success; William referred the petition to the Minister of Commerce and expressed the view that such patriotism deserved recognition. Nevertheless, the indomitable von der Heydt proposed to reject the application, whereupon William overruled the Minister and granted an extension of seven years for the patent "in recognition of the patriotic spirit so consistently and frequently displayed by Commercial Councillor Alfred Krupp of Essen, more especially in respect to his sacrifice of the profits which he might have derived from the sale of guns to foreign countries." Although this statement was hardly true in view of the orders already carried out for France, England, Russia, the Netherlands, and Egypt, still it served its purpose in enabling Krupp to win the fight over the wheel tire patent and his pretentions to patriotism were destined to bring him rich rewards in the future.

Krupp's foundry grew at a rate amazing even for a Ruhr district undertaking; by 1857 the employees numbered over 1000, in 1861 2000, and in 1865 8000. Vast extensions made to the works scarcely sufficed to deal with the rapidly growing output. In barely six years three machine shops, two gun shops, a tire rolling mill, a rail rolling mill, a wheelwrights' shop, an axle turnery, a gun hammer house, a plate rolling mill, and a boiler shop had sprung up round the original plant.

Although the business appeared to be so flourishing, there was surprising technical sterility in the Krupp works. They were reduced to using the inventions and research work of others to keep up to date in their production of steel. Having heard about the puddling process from the Zapp works in Ründeroth, Krupp proceeded to exploit it by the simple expedient of winning some of the Zapp employees over to his own employ. Henry Bessemer's great invention for making steel from iron scrap by the injection of air into the converter now threatened to outclass Krupp's method of smelting in crucibles, by which much of the iron was wasted. Through his London representative Longsdon, whose brother was a partner of Bessemer's, Alfred Krupp got into touch with the great engineer and acquired a German licence from him at great cost. The new process was treated with the customary secretiveness at the Essen foundry and the men were not allowed to know anything about the nature of the work to be carried out in the new building erected for the purpose. The converters were termed "Ratten" the product "C-steel." When it transpired that Bessemer steel could not replace crucible steel for high-grade products, Krupp did not take up his monopoly rights.

The introduction of the Bessemer process developed the Krupp works from a specialized factory into an ironworks. They expanded vertically at the same time by the acquisition of some fifty beds of iron ore in the Lahn district for half a million marks, and, in order to be independent of outside coal supplies, Krupp leased the "Count Beust" colliery east of the city. This expansion called for a reconstruction of the inside organization and the first attempt at collective control was made by including Pieper and Wiegand on the board of management. Sölling died in 1859 and his heirs' interest in the business was paid out.

The expansion of these years was due to the gun

business, for which Prussia's heavy orders gave the start. Others followed; Belgium, threatened by Napoleon, rearmed her field artillery with Krupp guns; Russia placed orders to the value of one and a half million thalers—a sensational sum for those days—Holland, Spain, Austria, Switzerland, Württemberg, Hanover and even Great Britain were numbered among the customers.

Krupp's personal position was now brilliant. After his visit to Essen, King William created him a privy councillor and later bestowed on him the Order of the Red Eagle with Oak Leaves—a distinction customarily reserved for a victorious Prussian general. After securing the Russian order, Krupp was openly referred to as "The Cannon King," and he proudly sent his wife a press cutting from Berlin, in which it was stated that the "roi des canons" had put up at an Unter den Linden hotel.

Borne on the wave of blood and iron, the little manufacturer had ridden high.

ON BOTH SIDES

It was about this time that Prussian political opinion began to take notice of this man of the Ruhr. The Progressives in the Chamber raised the question of his virtual monopoly and demanded that other firms should receive equal treatment. The Cabinet was in conflict with the Chamber but was compelled to give a promise to consider competitors in future.

This was a bombshell for Essen, and Krupp commissioned Meyer to protest to Roon about it. But the Government proceeded to place trial orders in Bochum and in other works in addition to those allotted to Krupp.

The latter hoped to defeat his rivals by exploiting monopoly in his material, but comparative tests indicated that there was little to choose between the various firms' products.

It was now a question of the lowest price, *i.e.*, a matter of calling for tenders in the usual way for the supply of armaments. Krupp was furious, as he considered the supply of cast steel guns an exclusive privilege of the Essen works and he wrote an indignant letter to Meyer, in which he threatened to supply guns with patent breech-blocks, hitherto only sold to Prussia, to any foreign customer who would pay for them. His hatred of anything which interfered with the interests of his business was boundless.

The Danish war of 1864 did not afford much scope for Krupp cannon, as out of the 110 guns comprising the Prussian artillery only thirty-eight were of cast steel. But on the conclusion of the war an order for 300 guns was placed with Krupp, although hesitation on the part of Roon had caused the King to intervene on Krupp's behalf.

In the autumn of that year Bismarck, coming from Biarritz, visited the foundry. He was the last guest in the garden house which, with its exotic park and many fountains, lay oddly enough amid the noise of the factory buildings. On November 2nd Krupp moved to his new residence, "Hillside," far away from the works on a height in the lovely Ruhr Valley surrounded by magnificent wooded grounds which he had secretly purchased. His relations with the works now underwent a fundamental change; he rarely set foot in them and as the telephone had not yet been invented, he had recourse to a tedious and time-wasting system of correspondence.

His chief concern was to overcome his competitors and recalcitrant Government officials. In 1865 Krupp attempted to purchase the Sayner ironworks from the Crown, as well as the adjacent Mühlhofen works and the Oberhammer and Horhausen collieries, for the sum of 400,000 thalers. The Minister of Commerce, von Itzenplitz, was about to conclude the deal when the secretly conducted negotiations came to the ears of Krupp's competitors. The Bochum Union violently attacked Itzenplitz and his policy, and compelled him to seek approval from the King and the Cabinet.

Krupp was in urgent need of the Sayner ore, owing to its outstanding qualities. Worried by the action of his competitors, he hurried to Bismarck, who met his arguments with the significantly sarcastic comment: "They may think that Count von Itzenplitz is getting 5000 thalers out of the deal." Bochum's attitude was threatening to wreck the proposed sale, so Krupp increased his offer by 100,000 thalers, thereby admitting that the terms originally envisaged with Itzenplitz were suspiciously low. The Bochum Union would not agree to the proposed sale to Krupp and also bid 500,000 thalers. Nevertheless the sale to Essen did take place. Bädeker states in his chronicle that "the Government were disposed to assist a firm which was becoming a national institution, as King William called it." Krupp's influence was now so strong that such action on the part of the Crown was regarded as a matter of course, as had already been the case with the public and personal business interests of his ancestors.

If the State would not assist voluntarily, a little gentle pressure could be exercised on it. Or a little diplomacy such as Krupp used with his patron Bismarck. There were still troubles in store for the works, as Krupp had failed to take the necessary steps to make the firm's rapid growth financially sound. At the end of 1864 the banker Deichmann warned him of the coming world shortage of capital and in the following year, when the purchases of Sayn and Nassau and the lease of the colliery took place, Essen's financial position became critical. The works were now valued at from 10,000,000 to 15,000,000 thalers, but required an equally great amount of ready money. At the moment Krupp needed several millions. The banks could not supply them, so there remained but one other source: Berlin, the Government. It was not going to be easy to persuade the Chancellor to assist, but Krupp was past-master in the art of wire-pulling and began by instructing Henry Haas to see what could be done in Paris in regard to raising funds, although he was to emphasize that it was only money that was wanted, and that there could be no question of control. He then went to Bismarck and pretended that he might have to permit foreign interests to acquire control of his firm. Krupp's own report of this interview to the works management is an unexcelled masterpiece of cynicism:

"He (Bismarck) was very upset over the matter and agreed to discuss it with the King and the Minister of War, but he stated that it would be hard to secure a decision without the approval of the Minister of Commerce. I treated the matter as a trifle and rubbed in the fact that if I availed myself of the offers of capital freely made to me in France, I might lose my future liberty of action, and the works pass under partial foreign control. I did not omit to say that I could sell out for 10 millions, any day."

Krupp deliberately deceived the Chancellor in regard to the possible French influence. He was not really thinking of any such terms, but his ruse proved successful and Bismarck's discussion with the King and pressure on Roon resulted in a huge order for coast defence and naval guns being placed with Krupp with a payment on account of 1,250,000 thalers. On his return to Essen Krupp nevertheless continued his Paris negotiations. The banking firm of Seillière granted him a credit of four millions without, of course, securing any influence in the management of the works. Bismarck's reception of this information is not recorded, but it may be assumed that he did not permit Krupp to see its effect on his plans.

In 1866, a depressing year for German industry, Krupp once more applied to the Government, this time for a grant of 2,000,000 thalers. He approached Bismarck, Roon, and the King and again talked of selling the works, but State funds were low and Roon's opposition stiffer. While negotiations were in progress the Minister of Commerce, von Bodelschwingh, resigned and was replaced by August von der Heydt. The latter offered a regular loan of 1,000,000 thalers from the National Loan Fund or from the State Bank, but Krupp declined to accept a bank loan as he feared imaginary dangers. The King had to intervene, remonstrated with Krupp about his distrust, and prevailed on him to accept a loan from the National Loan Fund. In his notes on the loan Krupp disgustedly referred to von der Heydt's action in approving the loan instead of making him a grant. He was incapable of displaying gratitude to the Government which was now financing him.

In the spring of 1866 intensive and competitive arming

on the part of the various German States increased Krupp's ordnance sales enormously. Baden and Württemberg each ordered thirty-two guns, Bavaria four, and Austria twenty-four. The object of these orders was no secret in either Essen or Berlin, as Prussia had hurriedly placed a gigantic order for 162 four-pounders, 250 six-pounders, and 115 twenty-four pounders.

The war party in Berlin was anxious to ascertain Krupp's attitude in regard to the possibility of guns supplied by him being soon used against his own country. In a letter dated April 9th, 1866, Roon undertook to intervene in Essen. He wrote Krupp asking him not to furnish guns to Austria without his own Government's approval or, alternatively, to state whether he was unable to give any such promise.

Krupp replied on the 13th to the effect that he could not enter into such an arrangement as it would be tantamount to a breach of contract. But that he would notify Berlin prior to the dispatch of any guns to Austria so that the authorities might make their own arrangements for intercepting them if necessary; in other words he did not want to lose the order and shifted the responsibility for the ultimate destination of the guns to his own Government.

He then went to Berlin to seek further orders in view of the imminence of war with Austria. While he was doing his utmost to influence the authorities to order more heavy ordnance from him, his representative in Vienna was doubtless urging the Austrian Government to place orders for guns in view of the preparations for war proceeding in Prussia.

Less than a fortnight later Prussian troops marched against Austria. In the sanguinary battles in Bohemia and in the Valley of the Main, Krupp guns were used by both sides for killing German soldiers. Their purveyor was incapable of feeling any scruples in regard to the artillery duel between the guns of Prussia and their "Austrian cousins," although reminded of that fact in a letter from General Voights-Rhetz. Krupp knew that nothing pays better than to sell to both sides simultaneously.

The troubles with Berlin continued. At the close of the war the Essen works were informed that several of their guns had burst while in use and had severely injured their gunners. This caused the partisans of bronze guns to blame the Krupp material. The Essen works were perturbed and took back the guns supplied the previous year in exchange for improved ones, while the King, in contrast to the distrustful Ordnance Board, supported Krupp. In November of the same year the works received an order, approved by a decree of the Royal Cabinet, for 700 new guns.

Attempts to gain prominence as designers also led to friction. The "taper breech block," an improvement in gun construction evolved by the Essen engineers, was submitted to the Ministry of War for test. The reply was the customary one given by bureaucrats to offers to teach them something new; the present pattern breech blocks were quite satisfactory. A significant postcript to the official letter curtly stated that the supply of ordnance, with vents as used by the Prussian field artillery, to foreign countries, was to be strictly prohibited. Krupp, indignant, was at a complete loss to understand the obvious reason for this prohibition.

His relations with the Russian authorities were far pleasanter owing to the absence of Parliamentary control and the marked effect of generosity to ministerial representatives and military officers. The goodwill of Grand Dukes was a little more expensive, but years of practice had enabled the firm to learn the method of securing it when armament orders were at stake. At Krupp's suggestion Russia placed a preliminary order for guns with taper breech blocks and when these were delivered the acceptance tests proved so satisfactory that a further order for 600 such guns followed, thereby establishing the supremacy of the Krupp breech block. Interesting suggestions of a technical nature were then made by Russian artillery experts, including proposals to reinforce the gun barrel by an outer tube, to minimize the danger of bursts. The idea, already known in connection with cast iron guns, was to shrink a white-hot outer tube on to the cold inner one. The subsequent evolution of this principle, emanating, as usual, from an outside source, resulted in the production of the celebrated "ringed" gun, which rendered the firm of Krupp so famous.

A recommendation by the Russian military attaché led to the engagement of the engineer Wilhelm Gross for the Essen works. Krupp chroniclers make little reference to him, but his employment had an immense influence on the most important branch of Krupp's manufactures, —construction of cannon.

Gross, a simple ordnance artificer in the Prussian artillery, who had already acquired an international reputation through his research work in ballistics, joined the drawing-office staff at Essen and quickly became its chief designer. He was the originator of a series of technical improvements during the next decade for which Krupp took the credit. Gross dealt with the technical side of all orders and quotations and his unrivalled knowledge of ordnance construction was almost equalled by his qualifications in the production of explosives. For all that, he never rose beyond a subordinate technical post as Alfred Krupp feared his abilities and distrusted him. Gross became a director of the works (a position which he retained till the close of the century) only after Alfred Krupp's death.

The conclusion of the war of 1866 was followed by an outbreak of cholera, brought home by demobilized troops. Owing to the crowded state and insanitary living conditions prevailing in Essen (the population of the city had doubled in the last six years), the epidemic raged there with particular virulence. Krupp, with his wife and child, left Hillside in a panic. He was taken ill en route, but refused to stay either in Sayn or in Koblenz, and hurried on to Nice while the plague took many hundred victims in Essen.

He spent the following three winters on the Riviera and when not confined to his bed by illness he led the luxurious life of a cannon king. He was courted by potentates and important personages of all kinds. Krupp reserved a special form of present for monarchs—a silver-mounted steel gun. Visitors of non-royal rank but occupying influential positions in their own countries also enjoyed the hospitality of the Château Peillon.

Essen meanwhile was ruled by an administrative board consisting of Pieper and Wiegand, who were later joined by the mining engineer Lorsbach and Judge Loerbrocks. The internal organization of the works now included twenty separate departments, each under its own manager. Krupp's relations to his own creation had undergone a complete change. For four decades he had guided it with strong hand through periods of doubt and difficulty. The man's will, so it seemed, was ever stronger than the possibilities of his business. But now, when he reached the mid-sixties, his own work grew beyond the man. The factory was too big for him to manage, and many of its departments were outside his technical knowledge and judgment. The market now absorbed war material with such avidity that Krupp's personal intervention was necessary only in some threatening reverse, or in the case of important new contracts. And then only to settle a dispute.

And while the armament king's illness in Nice was turning him into a nerve-ridden old man the works continued to outgrow him. It had worn him out. He lay by the wayside, but the factory grew and grew.

It required a particularly strong impulse to recall him to the whirlpool of business life. In 1867-68 the question of arms for the growing navy of the North German Confederation called for a decision. Berlin opinion in some quarters inclined to the light British muzzle loaders, which Armstrong had sold all over the world and the design of which rested on long years of British naval experience, while other opinion called for Krupp cast steel breech loaders. Even bronze guns had their advocates. There was more than a single item of business at issue here: the entire armament policy of the young German Navy was at stake. This brought Krupp home from Nice and he hurried to Berlin to use his influence with Roon and the King to defeat his British competitors. Although he had recently attempted to use the good offices of Crown Prince Frederick to help sell his own wares in England, he reproached the German naval authorities with a lack of national pride.

As the gunnery experts were unable to agree, comparative firing tests of Armstrong and Krupp guns were arranged in the first duel between the rival armament makers. The experiments were conducted on the Tegel range near Berlin in the presence of the King, Moltke, Bismarck, the Ordnance Board, and the Admiralty. The results were unfavourable for Essen, the armour plates bombarded by their guns remained almost undamaged. Krupp's engineers blamed the Prussian powder, which, they declared, burned too quickly for use in such heavy guns, thereby failing to create the necessary muzzle energy. Armstrong had used their own powder and their own gun crews. However, there was no glossing over the failure and Krupp's rivals triumphed. The Admiralty proposed immediately the equipping of three ironclads with forty-one heavy guns of British make.

The news of the Tegel failure reached Krupp in St. Petersburg, where he was attending another important gunnery trial. He induced the Tsar to send General Mayevski to Berlin to report the favourable result of the tests on the Neva. As a result King William once more intervened on behalf of his protégé and ordered a further series of comparative firing tests carried out under the conditions suggested by Essen. These tests were, however, a pure formality and on September 3rd, 1868, the King, acting on the ostensible advice of the Minister, directed that all guns for the three ironclads then building should be ordered in Essen without awaiting the results of any further trials. The first round of the fight for a Krupp monopoly of German naval ordnance was won and the sale of British guns to Germany finally made impossible. The appeal of the arms manufacturer to his own country's patriotism proved successful.

Krupp was still regarded as an outsider by his industrial neighbours and colleagues. He remained consistently individualist even during the first groping steps towards industrial co-operation. He violently attacked the "Rail Convention"—an association of manufacturers amongst whom the Ministry of Railways apportioned contracts. Although the Essen works with their immense resources secured the lion's share, Krupp denounced "the accursed association" in scathing terms.

The management of the Essen works suffered a severe blow by the death of Albert Pieper at an early age, due to overwork. It took ten years to replace him satisfactorily.

The immense growth of the firm's business, involving a turnover of many millions, led to the creation of a special "Personal Propaganda Fund," which Krupp cleverly utilized for the furthering of his firm's interests in Russia. He dispensed princely hospitality at his palatial residence in the Ruhr Valley. His guests included the Princes Charles and Alexander of Prussia with their families and retinues, the Crown Prince of Sweden, a son of the Emperor of Japan, the Russian General Todleben and many other important personages. Krupp knew exactly how to handle them and his generous gifts were purposely designed to remind them of his friendship. He presented teams of thoroughbred horses to private persons and the most modern heavy guns to monarchs. His expectations of the favours these gifts might bring were usually realized, especially in the case of Prussia and Russia. But lack of response on the part of the Duke of Brunswick and the King of Hanover caused him visible irritation.

OFFICER OF THE LEGION OF HONOUR

Alfred Krupp had always looked beyond the frontiers of his own country for business. In his young days he had travelled through France, England, and Austria and sent representatives to Russia and Turkey. He would do business anywhere if there was good money to be made, and did not lose this internationalism of profit when he became the great armament industrialist. He had once sold files and rollers in all countries, now he was ready to sell his guns to as many Governments and armies as possible. Business is business! The arms industry, like all other trades, was run for profit without any silly sentiment about patriotism.

It is strange that such a perfectly clear understanding of Krupp's mentality should have allowed numerous legends about his dealings with France under Napoleon III to arise.

The actual facts are that Krupp began to woo the French authorities as far back as 1843 when he dispatched two of his earliest cast steel musket-barrels to Marshal Soult, the War Minister of Louis Philippe, together with samples of steel cuirasses. "If these prove acceptable, there are hopes of securing the whole of the army contracts," he wrote optimistically. But the authorities in Paris were cautious and placed only a few trial orders in Essen. A year later Krupp begged of the Prussian Minister von Rother an introduction to James von Rothschild in Paris, that he might approach the French Mint and War Ministry. A somewhat singular request in view of the wave of anti-French sentiment sweeping through Germany at that time! But such things did not worry Krupp. Rothschild supported him energetically and awakened his hopes; he appointed a resident agent in Paris with showrooms for his wares.

In the year of the revolution he directed that the contents of the warehouse were to be well insured or placed in the custody of friends. "If only peace could be restored in France and normal, lucrative business for us resumed!"

When the first cast steel gun was completed in the autumn of 1847, Krupp was of two minds as to whether he would offer it first in Berlin or Paris. A promise by the Prussian War Minister caused him to decide in favour of Berlin. But when, after the Paris Exposition of 1855, Napoleon III created him a *Chevalier de la Légion d'Honneur*, he devoted himself with zeal to furthering business with France. Paris now showed some interest in Krupp products, and after successful trials the French authorities ordered two shell guns and then six steel billets for gun tubes in Essen. Following satisfactory tests of Krupp howitzers, General Morin contracted for the supply of 300 12-pdrs. The Essen works were, of course, quite ready to deliver these guns, but the French Government cancelled Morin's order on the plea of a temporary financial crisis. The real reason for their action was the outcry raised by French industrialists, more especially the firm of Schneider-Creusot who themselves had been making heavy cannon for the past two years.

It was just after the Paris Exposition that the idea of establishing a Krupp factory in France was first mooted. Krupp was interested in the scheme and was ready to accept it. The financial group which promoted it was backed by the "Crédit Mobilier," and as usual Krupp's inherent distrust of banks led to difficulties; and when the French group, inspired by Schneider, declined to file patents on Krupp's products, he withdrew. He was not, in principle, opposed to the scheme, but merely objected to what he regarded as unfavourable terms. Nevertheless his disappointment did not prevent his approaching the French War Minister in connection with the supply of prismatic powder, which the latter had been vainly seeking to obtain and which had been produced at Essen for some time past.

In 1867 relations between Prussia and France were strained to breaking point. Napoleon's noisy demands for the possession of Luxemburg rendered it increasingly difficult for Bismarck to continue putting him off with empty promises. Krupp appreciated fully the gravity of the situation, but this did not deter him from displaying an impressive array of his guns at the second Paris Exposition, opening in April of that year. The display was intended as a special appeal to Napoleon, himself a gun expert. The central exhibit on the Krupp stand was a huge built-up gun with a bore of fourteen inches. It took fourteen months to construct, had a weight of 100,000 pounds and cost as many thalers. Once more the works were awarded a "Grand Prix" and on June 30th Napoleon created the Essen ordnance maker an *Officier de la Légion d'Honneur*, a distinction which the latter gratefully accepted, disguising his disappointment at his failure to secure any contracts. His Paris friends, who had their own armament industry to think of, gave him plenty of verbal assurances but no written promises of future business.

In order to make sure that this omission was not inspired by the negative attitude of the French military bureaucracy, Krupp wondered if he might not be able to enlist the support of the Emperor as he had that of William in Berlin. He therefore addressed a letter to Napoleon dated January 23rd, 1868, and presented by Haas, in which he submitted particulars of gunnery tests carried out under the auspices of the Russian and Prussian military authorities and suggested that the Emperor, with his personal expert knowledge of gunnery, would doubtless be interested in them.

The humble terms in which the letter was couched should not be misunderstood. They were obviously intended to impress the Emperor with the merits of Krupp guns.

General Le Boeuf, President of the French Ordnance Board (the French press called him a relative of Schneider's) carefully studied the two trial reports appended to the letter and rejected the implied proposal. The material of the cast steel guns lacked uniformity and even in Prussia the question of reverting to bronze guns was under consideration. Moreover the French ordnance works were stated to be every bit as advanced as the Essen manufactory. The Krupp file at the French War Ministry contains the cogent entry: "Nothing doing. File March 11th, 1868." That was the end of the matter. After the fall of the Empire this entry was discovered included in the sensational publication: *Documents authentiques annotés. Les papiers secrets du Second Empire.*

The editor commented on Le Boeuf's entry reproachfully: "The general policy appears to have been to reject anything not produced, constructed, and tested in France. This was scarcely to be wondered at, when one remembers that the President of the Paris Chamber of

Commerce at that time happened to be . . . Eugène Schneider."

Notwithstanding the rebuff, Krupp was not so easily put off and he made a further attempt, this time by a direct approach to Napoleon to whom he sent a catalogue of Krupp guns with a covering letter dated April 29th, 1868, in which he respectfully begged the Emperor to look at the drawings and particulars of the cast steel guns described in the book.

The result of this personal letter was equally barren. The Emperor's reply was gracious; he thanked Krupp for the book, in the contents of which he displayed great interest and he expressed his hope that the affairs of the Essen works would continue to prosper, but . . . nothing more!

Fine words, but no orders. Krupp was still hoping for some on the very eve of the Franco-German War. Both the Emperor's letters prove this.

Krupp, the holder of the Legion of Honour, the persistent canvasser for French contracts and the writer of humble and respectful letters to the hereditary foe of Germany, appears in the light of a somewhat different character from that of the patriotic German arms manufacturer rejecting the offers of the intriguing Emperor, depicted to us by Krupp biographers. It was not thanks to him that the guns which sprayed death at the soldiers of Germany on the heights of Spichern and from the forts of Metz and Sedan were not of Krupp make. He did his utmost to introduce them into the French artillery although he was cold shouldered by Paris.

The patriotism of a cannon king is a fickle passion. It wanes in the shadow of any interference with business profits, but waxes in the sunshine of fat contracts from his native country.

These came in in the summer of 1870. The declaration of war, July 19th, was the start of prosperity for Krupp. On the very day of mobilization he offered guns to the value of 1,000,000 thalers as his contribution to a "war tax." Roon knew the Krupp generosity, which was invariably displayed with an eye to the main chance, so he courteously declined the offer. Krupp renewed his offer, this time suggesting that the guns actually in stock should be made available as the contribution of the firm, but once more Roon gave an evasive reply. Krupp continued to bombard the Ministry with further proposals, but as the King was now inaccessible they were ignored and Krupp had, perforce, to content himself with war contracts apportioned to him in the usual way. He might well be content, as the activity prevailing at the works was unparalleled, the Prussian Government contracts for guns doubling those of 1869. Berlin was, for the first time, Krupp's best customer, and the number of men employed at the works in the first year of the war increased by 3000.

But even then Krupp had no thought of neglecting his foreign markets. He remained a detached business man, even amid the prevailing war psychosis. When Berlin requested him to hand over guns under construction for Russia, he replied that he must first obtain the approval of his clients. He must safeguard his commercial honour, but was ready to question St. Petersburg. When, in the early days of the war, there was a possibility of a French advance across the Rhine and Meyer proposed to arm the workers with rifles, Krupp immediately rejected the suggestion: "It would be a stupid thing to do.

If the French come to Essen, we'll greet them with roast veal and red wine, or they're likely to shoot up our factory."

The multi-millionaire himself did not come into intimate contact with the war. He remained at Essen and devoted himself to a purely private matter—the building of a castle at Hügel to replace the manor house in the Ruhr Valley. Building operations began in the spring of 1870 from designs which he himself, as an amateur architect, greatly improved. The building material came from limestone quarries near Chantilly and was delivered to Essen by a firm of French stonemasons. When war broke out and it became necessary to divert the whole of the country's resources to it, Krupp declined to suspend building operations: "Proceed at all costs, even if we have to take men from the works," he ordered.

Henry Haas stayed on in Paris after the outbreak of war, and while there was desperate fighting at the front and Krupp was tendering for war material in Berlin, truckloads of French limestone continued to reach Essen via Belgium, while rollers and wheel tires were dispatched to French customers via England. Only the growing bitterness in Paris, leading to the expulsion of Henry Haas, put an end to these sales behind the enemy front. However, several of the French stonemasons remained at work at Hügel and Krupp insisted on their being well treated. He never allowed his private interests to be affected by national animosities.

As an arms manufacturer he did all he could to intensify the conduct of the war. He did not share the misgivings of court circles regarding a bombardment of Paris. He placed his 2000-pounder, long since obsolete, at the disposal of the army authorities for bombarding the Tuileries and he designed giant mortars and siege guns to hurl 1000-lb. shells into Paris. Roon declined these offers. The fire-eating contractor did succeed in obtaining an order for one high-angle gun to shoot down the balloons sent up in Paris, but the army authorities cancelled a contract for nineteen further guns of this pattern. Meanwhile the war in France came to an end. Krupp biographers greatly exaggerate the part played by Krupp guns, as only a portion of the Prussian artillery was equipped with them. Despite the success of the French quick-firing *mitrailleuses*, the superiority of German artillery was overwhelming. The actual number of French guns was thirty per cent. less, and they were mostly obsolete bronze ones which the French army had shortsightedly insisted on retaining.

The reaping of the financial war harvest continued unabated during the next years of peace. As in the sixties, there was a huge increase in the number of men employed at the works. In 1871 they numbered 10,000, in 1872 14,000, and in 1873 16,000. The German victory was a splendid advertisement for the Krupp gun and the next decade saw Krupp armaments commanding the world's markets in a manner never again to be equalled. Austria, Turkey, Egypt, China, Japan, Brazil, and Chile ordered Krupp guns, and the number turned out in 1873 exceeded the war output by 2000. Huge contracts from Russia were pending and Krupp did not give up hope of France. Haas returned to Paris as soon as peace was signed, and his first report noted declining animosity against Germany and expressed the hope of securing good business, despite the abhorrence with which the name of Krupp was regarded by good French patriots.

"THE HOME OF HIS ANCESTORS"

As Nice was still inaccessible to Krupp in the autumn of 1871, he chose instead the fashionable English resort Torquay, in Devonshire. His six months here eventually became of great importance to him. His intention had been to rest. But, scarcely arrived in Torquay, the Cannon King was caught up in a strange passion of activity, and spent days in letter writing.

He had suddenly realized that the establishment of a German Empire had brought a new era. His steel factory was now so huge that there could be no question of leaving its future direction to vague tradition if it were to continue to prosper. And behind the daily difficulties there arose a more important question. Krupp's fine scent for danger warned him that a new spirit had arisen among the masses, a spirit in which he, the patriarchal factory owner, saw merely blind greed. He wished to work against both these things, against confusion in the factory, and against the new social danger.

His imaginary fears were due to alarming news from Essen. A senior employee had left the works, violating his service agreement, and Krupp was indignant; he proposed to pursue the faithless deserter by means of legal action. A cast steel main shaft of a British steamer had fractured; it was, of course, the fault of her engineers, who allowed it to run hot. They had to be dealt with according to their deserts. The management of the works had long ceased to take the communications from Torquay seriously—they saw in them only the reflection of moods and fancies.

But the administration overlooked one important and

ever-recurring point: the demand for "Standing Orders," which Krupp wished to lay down paragraph by paragraph for the instruction and guidance of every single employee in the immense and elaborate works. He wished to take every possibility into consideration that there might be no stoppage of work or any conflict as to responsibility. "I must feel certain that, within 25 or 50 years, there can be no disposition or arrangements made at the instigation of any sort of malice." The senior member of the administration of the works, Ernst Eichhoff, a brother of Bertha Krupp, offered passive resistance to the demands from Torquay, apparently underestimating the stubbornness of his brother-in-law, or possibly feeling himself unequal to the task. Convinced at last that his threats and demands had fallen on deaf ears Krupp entrusted a Berlin lawyer, Sophus Goose, with the preparation and execution of the "Standing Orders." For the next ten years, Goose became leading executive in Essen. Krupp had always chosen administrators and representing executives from outside the works. In this he remained the patrician, desiring no curtailment of the distance between himself and his technical staff. But at the moment, he seemed to find that, outwardly at least, the opposite course might be best.

The remains of the old residence originally inhabited by Friedrich Krupp and intended for the use of the works manager were still visible amid the workshops. The house had sunk considerably because of the mine galleries under it and was in a ruinous condition. None of the furniture or fittings had been preserved and the less decayed portions of the building were used as a store house. Alfred Krupp now conceived the idea of restoring this "original home" of the family, in view of the importance it might assume in a time of social unrest as portraying the humble origin of the multi-millionaire industrialist, a "self-made man of the people."

Thereupon Krupp sent elaborate and exhaustive instructions for the complete restoration and refurnishing of his "ancestral home" to a semblance of its appearance thirty years earlier. The improved residence built alongside it and used by the Krupps for many years was demolished to emphasize the small and humble appearance of the original house.

His orders to have the interior redecorated and furnished "exactly as it once was" could not be carried out because even he himself could not remember how it had looked in 1844. The thing that now grew up under the uncertain memory of an old man was the deceptive copy of a long since decayed and vanished tradition. It was merely a bit of theatre scenery but it sufficed for Krupp. He gave orders that the little house should not be used for any business purpose but should remain as long as the factory stood as a "monument, as the origin of the great Works." It soon became a valuable bit of publicity for counteracting the covetousness of lesser folk.

After a short visit to London, where Longsdon called his attention to the newly discovered Spanish ore deposits, Krupp returned to Essen. A social crisis there claimed his immediate attention.

The Franco-German War had an alarming sequel in the Paris Commune; for the first time in history, the industrial working class took direct action and their struggles and deaths found a ready echo. Bismarck admitted that it gave him his first sleepless night. The German workers also began to show symptoms of unrest. As they awoke from the war psychosis they began to realize that a rise in living costs, and speculation in rents were making conditions worse and worse for them. Feeling ran particularly high in the Ruhr district, where the price of coal rose rapidly while wages and conditions of labour remained unchanged. When the colliery owners declined to negotiate, a coal strike broke out in June, 1872, which, after the intervention of the police, spread over the whole district. But after six weeks of unparalleled privations the miners were forced to give in unconditionally. They had lost their first battle through lack of organization, but they had fought it manfully and the more enlightened industrialists recognized that the voice of Labour could no longer be ignored.

Alfred Krupp regarded this development of the labour problem with dismay. His observations of the more advanced conditions prevailing in England, where the trade unions had already acquired a certain measure of influence, had filled him with horror. As a later historian puts it: "The great industrialists of the Rhine and the Ruhr were revolutionaries in all fields of technic, of trade development and finance. They broke through the old forms of production, built huge plants of hitherto unknown capacity, creating employment and a living for the rapidly increasing population so that the latter no longer were compelled to emigrate. But at the same time these great manufacturers believed that among their workers in factory, mine, and mill, in the growing cities, the same point of view, the same sense of duty and obligation could be retained that once held good for the smaller communities."

Krupp's behaviour during this time revealed him as a thorough-paced reactionary. He refused to listen to any appeal from his employees and openly said that he would rather blow up his whole works than submit to pressure from Labour and threat of strikes. As his employees were beginning to claim the right to form trade unions, Krupp, convinced that the movement had been fomented by agitators, attempted to arrest the march of progress by means of a paternal admonishment. On July 24th, 1872, he issued his famous appeal "To the workers of the steel foundry," which was to be the Magna Charta of industrial oligarchy for a decade to come. After a series of long-winded adjurations he summarized the relations between the employer and the workers in the following classical terms:—

"I expect and demand complete trust, refuse to entertain any unjustifiable claims, and will continue to remedy all legitimate grievances, but hereby invite all persons who are not satisfied with these conditions to hand in their notice, rather than wait for me to dismiss them, and thereby to leave my works in a lawful manner to make way for others, assuring them that I am and will remain master of my own house and business."

The trend of Essen's political development was the answer to this creed of an autocrat. Every election showed an increase in the number of votes for Labour, which finally won a majority.

The end of the Franco-German War eased the international political situation and saw the beginning of that Balance of Power in Europe which was to form a precarious safeguard for peace for the next few decades. But any who imagined that this marked the beginning of seven lean years for the armament industry, did not reckon with the inventive perseverance of these gentry. The nations had hitherto armed for war, they had now to arm to preserve peace. Who could tell whether it would last? Krupp was quite certain that France would soon be ready for another war and the only protection against this risk was the immediate rearmament of the artillery. The business significance of this admission did not seem to occur to Krupp. The armament maker's patriotic left hand never knew what the right hand was taking in in the way of business.

While Germany was still recovering from the effects of demobilization and the effort to revert to normal conditions, the indefatigable Krupp began his campaign for rearmament. As usual he had recourse to backstair tactics at court. Correspondence with General Voights-Rhetz and audiences with Prince Charles and the Crown Prince prepared the ground for the approach to Roon, to whom he made an extraordinary proposal. He would furnish the entire ordnance requirements of the army, whether for 1000 or 2000 guns, with the utmost dispatch and if payment could not be arranged for out of the French War Indemnity, he would waive his claim to it. In fact he would expect no payment at all if his beloved country was involved in another war, as he was ready to make any financial sacrifice rather than run the risk of its artillery being inferior to that of its enemies.

Roon, a Prussian of the old school, with an inborn aversion to the business instincts of the rising *bourgeoisie*, rejected the offer of the Krupp "sacrifice" out of hand and his letter concluded with the following malicious thrust: "Although I will refrain from a detailed examination of these proposals, I cannot withhold my admiration at the ease with which you dispose of your own financial interests in them."

Krupp realized Roon's meaning. His marginal entry

on the letter is "Sarcasm?" In a letter to the Emperor he made the following statement, obviously directed against the Minister of War: "An excess of zeal in the discharge of duty may be regarded as arrogance, but I object to any such imputation of my actions."

Whether it was zeal in the discharge of duty or in the securing of contracts the chance was too good to miss. Debates following on the experience of the last three wars indicated that there was a need for an improved type of field gun. This might involve contracts running into millions and it made Krupp's mouth water. Essen proposed a 3.2-inch gun with copper rifling. The conflict between Krupp and the Ordnance Board was not confined to technical matters only. He was not satisfied with the slow rate at which the generals proposed to carry out the rearmament. He wanted to complete in a few months "what dilatory red-tape procedure normally takes several years to do"—as one of his biographers writes.

From his Berlin headquarters, Hotel Royal, Krupp embarked on an underground campaign of memoranda, audiences with the Emperor, and "friendly intermediaries." His great triumph came when he succeeded in winning over Bismarck himself, who promised his support.

Krupp's special hatred was directed against Chief of the War Department, Lieut.-Colonel Willerding, and Inspector-General of Artillery, Hindersin, who joined Roon in opposing Krupp's efforts to secure a monopoly. Then began an underground persecution of both officers. Krupp followed the Emperor to Ems, where he succeeded in prejudicing him against the Ordnance Board. Indignantly he complained that they refused to let one of his engineers attend experimental firing tests. And he actually induced the Emperor to remove Willerding from his presidency of the Board by alarming the old gentleman with fantastic hints of a French invasion of Western Germany, and threats to sell his business.

Krupp's continued campaign finally enabled him to carry out at least a part of his scheme. He secured a few preliminary orders, followed by the main contract in January, 1874—providing for 350 complete field guns, over 2000 gun barrels and the necessary gun carriages and axles. Even his steel foundry, accustomed as it now was to huge orders, had never yet received one of so many millions. The Cannon King's ruthless and unscrupulous campaign had borne rich fruit; Essen had captured Berlin and nobody there would now dare to oppose him.

After unending delays and disputes the castle in the Ruhr Valley was at last completed. The industrial Crœsus gave it the modest name of "Villa Hügel" but the building was typical of the uncultured parvenu of the period. It was a cold and cheerless house, lacking comfortable rooms, pictures, statuary, and decoration. These things made no appeal to the totally inartistic owner, whose fear of fire even caused him to omit a library. The huge bare rooms were most depressing in the rainy climate of Essen and no fresh air penetrated them, as the elaborate ventilating system, of which the hypochondriac Krupp was extremely proud, was invariably out of order. His study windows overlooked the stables—he insisted on this arrangement as the smell of the stables stimulated him and he was, moreover, able to keep an eye on the servants.

Despite the brilliance of Krupp's external life at this

height of his successful career, his mind became increasingly depressed and he distrusted even his closest relatives and co-workers. He quarrelled continually with his wife Bertha, whose constant stream of guests robbed him of the peace and quiet he urgently needed. Alfred Krupp also fell out with his brother Hermann, the successful manufacturer in Berndorf, whom he wrongfully accused of spying out technical secrets in Essen for the benefit of a new Austrian works. Ida and Fritz Krupp kept away from Essen, but forwarded fresh claims concerning their heritage. Although barely in the sixties, the cannon king had become a querulous old man, whom his doctor found hard to persuade to leave his bed—even when travelling.

One peculiar complex he now displayed was his aversion to the Works, which he actually avoided. He even refused to visit them in the intervals which he spent at Villa Hügel between long absences in various parts of Europe.

THE INDUSTRIAL CRISIS

An orgy of wild speculation swept through Germany's economic structure in the first year of the new Empire which had broken open the portals of a world-market with its victorious bayonets. The unexpectedly rapid inflow of French money in thousands of millions was expended on armaments and subsidies, thereby producing an alarming increase of the capital to invest. Banks, factories, railway and mining undertakings sprang up overnight. The average annual number of limited liability companies registered in Prussia between 1851 and 1870 had been only 18. In 1871 there were over 200 such registrations, and 500 in 1872.

"Get rich quick" was the motto of the day. Even the great Bismarck dabbled in business which was discreetly handled for him by the banker Bleichröder. Fantastic earnings made Henry Strousberg, once a small commission agent, into a "railway king" with a palace on the Wilhelmstrasse. Members of Parliament, cabinet ministers, plebeians and patricians joined in the wild scramble for huge profits by coaxing money for worthless shares out of the pockets of humble folk.

Alfred Krupp and his Works were, of course, affected by the universal fever, and although he would not issue shares himself, he readily plunged into the wildest speculation. In spite of the immense growth of the foundry during the preceding decade, it now expanded at a terrific rate. New workshops sprang up and the number of furnaces, boilers, and steam engines was vastly increased, while the number of employees rose by over 5000 between 1870 and 1873. Apart from war material the industrial boom brought orders for thousands of miles of rails, thus almost doubling the total output of iron and steel from year to year.

Much of the new development took place outside Essen itself. There was an apparent shortage of raw materials with a consequent rise in prices, starting competitive buying of coal mines and ore deposits. Krupp, not to be outdone, joined in and paid fabulous prices, and, as with Hugo Stinnes half a century later, his operations were the talk of the country and upset the markets.

Krupp tried to corner all sources that might give him independence in raw materials. He bought over 300 ore deposits in various parts of West Germany and acquired a holding in the Orconera Iron Company, which owned large concessions in the valuable ore fields of Bilbao, Northern Spain. Although the coal supply for the Essen works was already assured through contracts with several collieries, the "Hanover" Mine was bought for four million marks. To crown all, Krupp completed his year's purchases by buying the Hermannshütte Ironworks at Neuwied and the Johanneshütte Ironworks in Duisberg, of which the last named alone included four blast furnaces. To facilitate the transport of the Spanish ore a small fleet of ships was built, with steamers docked at Flushing and, later, at Rotterdam.

The financial liability of the firm was rising to astronomical heights and the capital value rose from fifty to eighty-eight millions in four years, notwithstanding substantial writing-down on various occasions. Krupp's craze for expansion knew no limits and he suggested to the administrative board that the firm should erect a large blast furnace at Segeroth near the steelworks. But before this scheme was realized the boom ended.

The crash came in 1873. It started with a financial upset in Vienna, which, affecting all stock markets in Europe, brought on an appalling crisis. There was an unending series of bankruptcies and most of the new shares bought at inflated prices proved worthless. Industrialists whose names were household words collapsed in the general crash.

As unsuspecting as his entire epoch, Krupp found himself facing the most serious crisis of his life. He had had ample warning. Ernst Eichhoff had called his attention to the huge indebtedness to the banks and to the gross overcapitalization of his firm.

Krupp's only reaction had been to make a trustful

appeal to Eichhoff's ability to see him through. He had not understood the warning. He continued to raise money as before, the sole difference being that instead of long-term loans he took short-term credits.

Funds began to run low at Essen as early as the summer of 1872, while mountains of raw materials had been secured. The foundry owner turned to his usual source of supply—Government aid. In July he went to see the Emperor in Ems in order to sound him as to the possibility of securing a State loan of five million thalers on account of future contracts—failing which, it might be necessary to depart from the principle of progress and sole ownership, hitherto in force. William, kind-hearted and obliging, promised his support, but opposition from the Minister of Finance wrecked the scheme. Meyer had meanwhile endeavoured to raise money from the banks, but Krupp's fear of the influence of Hansemann, powerful chief of the Diskontogesellschaft, prevented anything being done in that direction.

The financial position soon became acute. Five millions no longer sufficed; the works now urgently required double that amount. Krupp went off to Italy on medical advice but he dispatched Meyer to see the Emperor with a desperate appeal for a State loan of ten millions. William asked innocently, "How is it that Mr. Krupp now requires such a lot of money? He asked for only four millions in Ems." Even his and Bismarck's influence did not help much, as the loans eventually approved could not suffice to cover the most pressing liabilities of the firm.

Meyer once more turned to the banks, negotiated with Bleichröder and the Maritime Bank. But the bankers had a shrewd idea of the position in Essen. When Meyer-

Cohn and Deichmann granted larger short-term credits, the banks warned them that caution was necessary owing to Krupp's mania for buying everything in sight.

In the spring of 1874, at the peak of the crisis, Krupp himself hurried to Berlin. It was now a case of prompt action if disaster was to be avoided, but Bismarck was ill and the aged Emperor unequal to the task of overcoming the opposition of the Minister of Finance single-handed.

Only the banks remained now. Alfred Krupp, who had always shown his distrust of them openly and conducted financial operations on a gigantic scale without their assistance, saw himself compelled to turn to them for help. What were their terms likely to be? Krupp's connection with the Government might enable him to avoid sharing the fate of so many other great manufacturers, who had been swallowed up by the banks. After prolonged negotiations a syndicate was formed under the auspices of the Government Maritime Bank, which offered to provide a loan of thirty million marks, repayable within ten years and with a mortgage on the entire works as security. Some of the terms were very harsh, and the banks insisted on the unwelcome Meyer joining the administrative board as their representative.

Krupp's angry disappointment unloaded itself on the Works. Salaries and wages were reduced by fifty per cent. within the next five years. As usual, the heaviest burdens fell on the weakest shoulders and the management intimated that any complaints would be regarded as justification for dismissal. Although Krupp himself was unaffected by these retrenchments, they were rigorously applied even to the senior executive staff who had to bear the responsibility for his mad policy. Outside auditors were brought in and their reports presented a gloomy picture of the financial position and of the basically unsound structure of the business. They made an extremely painful impression on Krupp, who declared himself to be "nearly out of his mind with worry and anxiety." All accounts were now examined with the utmost care and a special audit department was created to exercise a drastic supervision of all expenditure.

All this did not suffice Krupp. He grew more and more suspicious and scornful of his fellow-men.

The worst of the slump was scarcely over, in 1875, when the Empire was shaken by a fresh political crisis. The reorganization of the French Army by MacMahon gave the jingoes of the German General Staff under Moltke their chance to urge a preventive war to crush France completely. An inspired press campaign added fresh fuel to the flames and the Berlin *Post,* the organ of the armour-plate manufacturer Stumm, published an alarmist article entitled "Is War in Sight?" As a climax to this agitation the German Ambassador in Paris called on the French Foreign Minister and warned him of the consequences of France's rearming.

That was a formal threat of war which threw all Europe into a panic. The British Government called upon Berlin to keep the peace, while the Austrian Emperor and the King of Italy met in Venice. The Front was closed against Germany and the final intervention of the Tsar, through his Foreign Minister Gorchakoff, forced the German war party to retreat, thereby administering a sharp rebuff to Bismarck.

Krupp was once more, by force of circumstances, compelled to choose between the interests of the State and his business. He turned to Austria, where for years past he had been hoping to secure an important contract. In 1872 the Austrians ordered an experimental gun in Essen of a type hitherto supplied only to Prussia and Russia. After firing tests the Vienna authorities asked for quotations for the supply of 2000 such guns and the amount owing for the experimental work. Krupp refused to make any charge for the latter, but expressed his hope that he would be favoured with their order. There was some delay in placing the order, however, and further tests were called for before making a decision between the rival merits of bronze guns of Austrian make and the steel guns from Essen. Krupp sent his best men, including Gross and Meyer, to convince the Austrian army authorities.

Unexpected opposition was now raised in Berlin, where the military authorities regarded the Prussian type of field gun as the special preserve of Germany and not available to foreign countries without their express permission. Furthermore, the design of the gun had been prepared by Essen with the assistance of army artillery experts. Krupp's retort to this was to take the offensive; he directed Gross and Richter to publish "A History of the Cast Steel Gun"—a technical work of necessarily limited appeal. The book greatly exaggerated the part the Essen works played in producing the gun, but succeeded in its provocative object. Berlin forthwith called upon Krupp not to publish the book openly nor to make it accessible to foreign powers.

The alarmed Emperor now took a hand. He heard to his dismay that the man who always talked like an out-and-out patriot proposed to sell guns of Prussia's special design to a power which had joined an anti-German coalition. William was particularly hurt to learn that it was Krupp who sought to rearm Austria.

The outcry against the arms manufacturer now became more serious, and in view of Krupp's negotiations with Austria the Minister of War prohibited the export of guns. This affected the life blood of the firm and just as eight years earlier, when von Roon insisted on imposing restrictions due to political considerations, Krupp proceeded to defend his rights with the utmost energy. He went to Berlin and sought an audience with the Emperor, to whom he explained that the German Government contracts alone could not keep him going, as he required orders for at least fifty millions during the next ten years to pay his overhead. He could not be expected to sell inferior products to foreign States, but in any case he was not proposing to sell guns of Prussian design to Austria, but rather Russian designed ones.

William was duly reassured and Krupp once more succeeded in allaying the misgivings of the monarch, who completely overlooked the main point, *i.e.*, that Krupp himself had really corroborated the main point of the charge against him, that of offering to supply arms to hostile States.

But even after the cessation of opposition in Berlin business with Austria fell through and all Krupp's personal efforts to back his local representative in Vienna were fruitless. He thereupon accused the Austrian army authorities of trying to steal his technical secrets and threatened to expose them publicly; but even this did not influence the War Ministry in Vienna in its decision to give their own manufacturers preference over Essen. They finally found a solution of their dilemma by adopting the "steel-bronze" invented by Major Uchatius and arranging for the construction of guns from this material in their own ordnance factories. To compensate the Essen works for their samples and the loss of the contract, they were given a payment of 160,000 florins.

The cannon king, however, insisted on vindicating his right to sell his wares to all and sundry although he expected his own government to order guns from him only. Calls for open tenders infuriated him and when he heard that the Berlin authorities were now about to invite a number of German firms to bid for the supply of guns, he addressed a vigorous protest to the Minister against infringement of his "rights." He did not possess any patent on the construction of the gun, the design of which was not even the property of his works. Artillery experts had collaborated with the factory in important details. But Krupp had found that he could do as he liked with foreign customers, so he saw no reason to treat his own Government differently.

He was sublimely unaware of the inconsistency of his position. He threatened to sell his works and emigrate if the Ministry should dare to entrust contracts to his competitors—"a most effective way of exercising pressure," as Krupp told his confidants.

His next audience with the Emperor—whom Krupp tried, as usual, to draw into the dispute—proved fruitless, as William declined to express any opinion about Krupp's real or imaginary grievances. The Ministry of War would not depart from their ruling; they reserved the right to accept any tender, preferably the lowest. But Krupp's violent campaign had, to a certain extent, won him a privileged position and although the present large contracts were not all placed with him, the time was coming, under his successor, when the monopoly for which Alfred Krupp had worked was to become a fact.

After the effects of the prolonged depression had passed, about the end of the seventies, the United States embarked on a vast programme of railway construction. The Essen works struggled upward again. Their exhibit at the "World's Fair" in Philadelphia was a 14-inch gun of sixty tons weight which was the central exhibit of the German section. Unfriendly comments on this warlike spirit in times of peace induced the administrative board of the works to call attention to other spheres of the firm's activities. Their stand at the Brussels Exposition displayed models of their workmen's houses, schools, canteens, and hospitals, with booklets describing their ideal welfare organization. This marked the beginning of a highly effective publicity campaign, representing the Cannon Firm as a support of all modern social welfare.

BLOODY INTERNATIONAL

What armament makers particularly like is to sell to both sides of the hostile front. And when this concerns "exotic" countries the profits are fourfold; one can sell them obsolete, and surplus or doubtful stock. "Chinese and Siamese can blow their enemies to bits well enough with these," said Krupp about a batch of obsolete guns. His deliveries to Eastern Asia were carried out on that principle. Huge orders from China and Japan preceded the war scare of 1874-75. Japan prepared for war with China, and China equipped the Taku Forts with Krupp heavy guns, which were noted with surprise by the visiting German warship *Ariadne*. Twenty years later Commander Lans of the *Iltis*, severely wounded himself, was to protest at the scandal of his men being blown to pieces by Krupp guns in the Taku Forts.

The representatives of the Essen ordnance works, in their search for new business, penetrated into the awakening cockpit of Europe—the Balkans. They had to proceed cautiously but still annoyingly enough for Bismarck's foreign policy, which was intent on preserving the friendship of Russia. As a preliminary to the disastrous Bagdad policy at the beginning of the next century, the first large Krupp contracts were now secured from the Sublime Porte. Turkey figured prominently amongst the customers of 1873, with extensive orders for field artillery, naval ordnance, and heavy guns for the Dardanelles defences and the forts on the Bosphorus. An additional Turkish order running into millions came as a godsend to the steelworks, then at their lowest ebb due to the slump. But its effect on Russia was disturbing. This early customer of Krupp had for years past been endeavouring to produce ordnance in her own works. But the slackening of Russo-Turkish relations and the huge Turkish contracts placed in Essen alarmed Russian military authorities who now pressed for a hastening of their own armament programme. Soon fresh Russian trial orders reached Essen. "Poor Turks, if Russia really gets those guns," said Krupp thoughtfully. He had learned from a remark dropped by William I that war between Russia and Turkey was a foregone conclusion. As St. Petersburg orders were still very small, Krupp's sympathy was all for the Turks who had paid him good money in plenty at a time when he needed it badly.

However, the trial orders from St. Petersburg were productive of something very substantial and in 1876 the Russian Government placed contracts with Krupp for the supply of 1800 light and heavy guns, valued at twenty million marks. In face of such a sum Krupp's sympathy for the poor client on the Bosphorus melted away and he was now full of enthusiasm for the Russians. He made them a present of his Philadelphia 1000-pounder, gave the Tsar and Grand Duke Michael exquisitely finished ceremonial guns and omitted to invite the Turks to attend gunnery trials at Meppen. The big contract won the victory.

Meanwhile, war clouds in the Balkans had burst. The Serbian and Montenegrin risings against Turkey were, in 1877, backed by the intervention of Russian and Roumanian armies which brought success. Krupp guns had already been used by both sides in the fighting between Serbia and Turkey. But the battle of Plevna showed their use by both combatants on a vast scale, the Turks having sixty and the allied Russians and Roumanians twice that number of Krupp guns. After long desperate fighting the Turks were defeated, their defeat being the first stage in the collapse of their power in the Balkans and of their virtual expulsion from Europe. Krupp was right in pitying "the poor Turks" when selling guns to their opponents.

He did not hesitate to admit supplying arms to both sides. In a memorandum which he distributed in 1879 to the British House of Commons, he boasted of satisfactory reports on Krupp guns from Russian, Turkish, and Roumanian sources during the Russo-Turkish war.

The experience gained in this war brought Krupp several new customers. Following orders from Spain, Italy, Denmark, Sweden, Portugal, and the Netherlands, Greece placed a large order for field guns and mountain artillery for immediate delivery. While other German manufacturers were still suffering from the full effects of the industrial depression, the Essen works were so busy that an increase of a full thousand men in the number of their employees proved necessary during the year 1877. The profits of Balkan war contracts paid off in full the loan negotiated five years earlier. The mortgage on the Works remained, but the total liabilities were reduced from thirty to twenty-two millions and repayment was spread over twenty years. This solved the most acute problem arising from the depression.

Changes in administrative personnel also took place as Krupp was still resentful of the recent financial crisis. The first to go was Richard Eichhoff, one of the old-timers who, like Ascherfeld before him, was inconveniently conscious of the fact that the success of the Works was largely due to his efforts. Lorsbach, Lörbrocks and Wiegand left the administrative board and were replaced by the mining engineer Ehrhardt, the accountant Gussmann and the merchant Cohnheim. In 1879 chairmanship of the Board passed to Councillor of Finance Dr. Hanns Jencke, who acted as Alfred Krupp's personal deputy and represented him in all internal and external business matters throughout several decades.

The year 1879 brought a change in German economic policy. Bismarck, backed by the National Liberal Party and working through Delbruck, Minister of Commerce, had carried on his free trade policy for eleven years. But the depression brought more and more opposition, especially on the part of the iron industry pressed hard by English competition. They had formed an alliance with the big land-owning interests and their aim was to force protective duties at the expense of the consumer. Krupp did a great deal of back-stage propaganda.

"One factory after the other will collapse. The iron works will be in the same condition as the ruined baronial estates. I face the fact that my business will suffer the same fate."

This was the gist of the material he sent to Meyer, who was to make an appeal to the Emperor.

At the same time negotiations were going on for Krupp subsidiaries in Russia, Austria, England, and even Japan.

In an audience with the Emperor, who, like Bismarck, had no head for figures, Krupp painted a picture of German industry in the darkest colours and succeeded in shaking his faith in the policy of economic liberalism. Bismarck needed the support of the Right for his anti-Socialist campaign and the Emperor still regretted the break with the Conservatives. A petition, signed by leading industrialists, including Krupp, for the abolition of free trade, came at an opportune time and served as its death-knell.

The brilliant work of Gross had enabled Krupp to realize his intention of designing guns as well as constructing them. For that he required an experimental artillery range of his own, as the space available for tests and experiments near the works had become totally inadequate. Apart from that, a private range would make him more independent of the official Ordnance Board.

The Ministry of War regarded Krupp's proposal with some suspicion and he therefore considered acquiring a range on foreign territory, in Russia or Holland. But finally he found exactly what he required in his own neighbourhood, and bought or leased a site at Dülmen in

Westphalia some thirty-five miles north of the works. Krupp thought the place ideal, but Gross, who had not been consulted prior to its acquisition, pointed out shortcomings. Krupp ignored him purposely, as he disliked his outspoken opinions. The retired Lieutenant Prehn, a coming man in ordnance work, was appointed to take charge of the range.

A few years later Dülmen proved much too small for tests of heavy guns and it became necessary to find another site. After some search a suitable site was found near Meppen, not far from Osnabrück, and the Essen works secured possession of it by 120 separate leases. The entire range was surrounded by observation posts controlling the three highways traversing it and warning all passers-by of the proximity of the range. Huge cranes on concrete emplacements were provided for handling the gigantic 17-inch guns and shell-proof shelters protected the crews from stray shot and splinters. The equipment of the range was on a larger scale than anything of a similar kind owned by a private armament firm or Government anywhere. The name of Krupp was heard again throughout the world.

The Meppen range became the scene of important gunnery tests, attended by visitors from all countries. Krupp flatly refused to allow Prussia any special privileges on his expensive range. In the summer of 1878 his visitors included twenty-seven artillery experts from twelve foreign countries, besides German ones, all of them being entertained by him and accommodated free of charge in his private hotel in Essen. After watching demonstrations of an armoured gun at Bredelar the visitors spent several days in inspecting the works. Then young Fritz Krupp escorted them to Meppen, where the latest patterns of heavy guns were demonstrated by actual tests.

An even greater "international shoot" took place in the autumn of 1879. It was attended by representatives of almost every country in the world and Krupp himself defrayed the cost of entertaining a hundred senior naval and military officers of twenty different countries. Guns of twelve different sizes were demonstrated at Meppen, including the 17-inch built-up cannon, the heaviest gun in the world.

The results of both these events were as expected. Orders came from all over the world. "What do a few thousand thalers for expenses matter?" said Krupp. During the tests it was noted that the attendance of British officers was larger than that of the German and that no French officers were present. This was not due to any patriotic sentiment on the part of Krupp, but to his innate distrust of the French and his failure to secure orders from them after years of unceasing efforts in Paris.

Otherwise the international collection of dealers in murder was complete and swarmed round the works and Hugel like vultures. Agents, Government representatives, accepted cadgers, and courted clients made a mixed assembly from all quarters of the earth. When the principals themselves attended, Krupp displayed the glamour of his industrial court, at which French, Italian, and English—but no German—were spoken. His visitors included Asiatic potentates, royalty, ministers, and generals from scores of different countries.

Records relating to this portion of the Krupp firm's activities were kept in the confidential archives, which included a secret section known as "the poison cupboard." One of Krupp's biographers (Berdrow) makes a guarded reference to it by recording a remark of an unnamed but highly placed personage—presumably Moltke or Roon—who, after a visit to the works and to Hügel, warned his subordinates against Essen: "Better not go there. Manufacturers are apt to get round people by good dinners, that their tongues may be loosened by wine," etc. The indignant Berdrow comments: "And he was one of the best!"

On Sedan Day in 1877 the Emperor honoured Hillside castle with his fourth visit. His large retinue included a dozen princes and generals. In spite of numerous vexations William still retained his interest in the Krupp works. The generally assumed "small investment" of the monarch (consistently denied by the Keeper of the Privy Purse) was merely a result not a cause of this interest, which was primarily due to William's taste for military matters. Considerate gifts from the industrial magnate served to remind him of his friendship in a pleasant manner. Among them were two beautifully decorated guns for the Imperial yacht "Hohenzollern."

Although Krupp disliked paintings, his private rooms were adorned with numerous portraits of German generals and foreign potentates and notabilities. He had sufficient motive for "eternal gratitude."

SOCIAL ANXIETIES

One of the most persistent legends about the firm of Krupp is that which credits them with good treatment of employees. Alfred Krupp's admirers insist that he was in advance of his time in regard to welfare work, sick and provident funds, housing and canteens for his work-people. But even a superficial study of actual achievements in this direction should serve to convince anybody that, far from being in advance of his time, or even up to date in regard to it, he frequently lagged behind in many ways.

The sick fund of the Works was already in existence in 1839, when the firm presented it with two thalers levied in fines. But regular subscriptions to the sick fund were apparently collected from the workers at the rate of a penny per man per week, while the sole contribution made by the firm was an annual payment of twenty thalers to the medical officer. This scanty provision for illness was already in force in the Rhineland thirty years earlier, having been established by the French during their period of occupation. The formal sick fund and pensions fund of the Krupp works were not established until 1855, *i.e.*, eighteen months after enactment of laws providing for such insurance. Even after this date we find that the firm, although contributing only half of the total statutory payments, exercised a despotic control over their administration. Contributors who left, or were dismissed from the works, had no claim whatever on the funds and frequently forfeited subscriptions paid for ten or more years.

The housing plan, second showpiece of Krupp "welfare work," came into being through the needs of Essen's industrial growth. At the beginning of the century the population numbered a bare 4000, and even as late as 1850 had risen only to about 10,000. Then came a sudden jump to 50,000 of whom 10,000 were employed at the steel works. Conditions of that day were described by one writer as "Unsanitary living conditions, abnormal rise in rents, exploitation of the workers by shopkeepers and saloons."

The cholera epidemics following the war with Austria showed the danger of these conditions. Other establishments might permit themselves a continually fluctuating working-force, but Krupp required, for quality product, a trained and experienced personnel.

As private housing development did not keep pace with the growth of the works, Krupp was forced to provide houses for his workpeople himself if he wished to retain their services permanently. That was the origin of the six housing schemes developed around Essen. All of them consisted of houses of the simplest, almost primitive type, which barely met the requirements of the humblest class of workers. The industrialist was opposed to any trace of "luxury"—his idea of housing the men being to build thousands of huts similar in type to those in which his own ancestors had lived. Old age alone prevented him from realizing this philanthropical scheme. Some of the housing schemes carried out by Alfred Krupp—*e.g.*, the Schederhof—are to this day the worst slums in Essen. As the payment of rents for the workers' dwellings was guaranteed by the works' pay office, it is not clear whether the "welfare" provided was for the workers or for the firm, which was hereby enabled to exercise a strict supervision of its employees.

Krupp also established his own hospital and a hotel, built schools, and opened a chain of bars and canteens all over Essen. The canteens retailed goods at ordinary prices and the discounts they gave were lower than those given by local co-operative stores. Moreover, the benefit of even these discounts was forfeited by workpeople leaving Krupp's employ during the year.

The Krupp Works had, therefore, developed into a "state within a state" as Jérôme Bonaparte called them on the occasion of an unsuccessful visit to Essen. The Krupp workman was under the control of his employer not only in the Works but in his home, during his spare time, and while doing his shopping. Admittedly, his standard of living was an improvement on that of many other workers in the district as Krupp had more vision than the short-sighted industrialists of an earlier generation. He cared for his men, because he recognized that any improvement in their living conditions would react a thousandfold on their productiveness and sense of responsibility. Still his actual expenditure on their welfare was grotesquely small when compared with the huge profits their work brought him, and his trifling outlay on their well-being pales into insignificance beside the princely endowments of Carnegie and Nobel.

Krupp's behaviour as chief of the factory was characterized by a complete absence of any trust in his men, and no trace of the much-boasted good relations between employer and employee was ever to be found in the Essen Works. The "Rules for work employees" issued by young Krupp in 1838 contain the following typical illustration of his mentality: "Any person found sulking or slacking will be dismissed, likewise those who are guilty of repeated negligence. Eye-servers must expect dismissal at the first opportunity and insolence will be punished by summary dismissal."

His views did not change with the passage of time, as is proved by hundreds of suggestions put forward by Krupp to the administrative board. Men of average intellect, devoted automatons, represented his ideal. "Ability is no equalizer, but if devoid of morality, is more dangerous than average intelligence." He would have liked to dress his men in uniform (he did, in all seriousness, suggest doing so) and to regulate their private lives that he might "provide the State with many faithful subjects and develop a special breed of workers for the factory." He strove to exercise constant supervision over them, to shield them from any outside influences, and enjoined his management to watch the men at all times. When he learned that certain notices put up by the works management had been torn down, he gave orders for a constant watch by thoroughly reliable men, regardless of expense, to detect the offenders, with a reward for every culprit arrested.

As Krupp grew older, he became even more autocratic. In 1877 he issued his famous "Words to the employees in my industrial establishments." He announced that socialism meant the rule of "the idle," "the dissolute," and "the incompetent," and threatened the very existence of the factory. He would not hear of any talk of the workers' political rights. "Serious concern with politics involves more time and thorough study of difficult questions than you can afford." Those who, nevertheless, insisted in dabbling in such dangerous matters, were sternly admonished in the following terms: "Every man must be responsible for the consequences of his own conduct. One does not cherish a viper in one's bosom and whoever is not wholeheartedly devoted to us and opposes our instructions cannot be allowed to remain in our employ." This was no empty threat: shortly after this pronouncement thirty men were dismissed for spreading socialist doctrines.

The relations between Krupp and his men were further strained by religious differences. The citizens of old

Essende at the close of the Middle Ages were, in consequence of their long feud with the local abbey, exclusively Protestant. The development of industry attracted swarms of young Westphalian peasants to the city and many of these took service with Krupp, thus bringing in a large number of Catholics, whose presence in the city gave rise to trouble. Alfred Krupp himself had no particular sense of religion, as is proved by his numerous writings and sayings. He adhered to the Protestant Church merely because it happened to be the State Church and as such better suited to his business needs. A patriarchal despot, he was in complete accord with the Berlin authorities' practice of treating the Catholic inhabitants of the west as second-class citizens, and of ruling the Rhineland through East Prussian officials like a colony acquired by conquest.

Religious differences did not arise until they touched on matters of social policy. In 1869 a Christian Labour Union was established in Essen under the auspices of the junior clergy, with its own newspaper, *The Essen News*, edited by Gerhard Stötzel, a former fitter and turner from the steelworks. His writing carried weight and his support of the strikers in the great coal strike earned him the enmity of Krupp, who accused him of libelling the management of the steelworks and of seeking to stir up trouble among the men. It made no difference whether the movement was Christian or Socialist—any form of Labour organization was a criminal conspiracy in Krupp's eyes.

The Reichstag elections gave rise to eternal bickerings between Krupp and his Catholic workers. Essen had, hitherto, elected Stötzel by an overwhelming majority. After the attempts on the life of the old Emperor, Bismarck dissolved Parliament and endeavoured to secure a majority by an appeal to the electorate. Numerous prominent industrialists were nominated as candidates in order to avoid expensive and cumbersome dealing through "friendly" Representatives. When a meeting of the local Government Party in Essen proposed to nominate Alfred Krupp as a candidate, the latter raised no objection. But at the elections held on July 28th, 1878, Stötzel was returned with 14,000 votes while Krupp only secured 13,000. The city electors, including many of his employees, rejected him by a substantial majority. Attempts have since been made to explain away this regrettable episode by excuses concerning doubts as to whether Krupp actually was standing for Parliament and the consequent confusion of many voters. But when the next elections were held in the autumn of 1881 it was noted that Krupp declined to stand again. He recommended Count von Moltke, a strong supporter of the big armament policy and a staunch friend of the arms manufacturers. Nevertheless, Stötzel was again elected.

Krupp's last encounter with the Catholics took place at the septennial elections of 1887, with the seven years' armament programme as an issue. On this occasion young Fritz Krupp stood in opposition to Stötzel, who was against the seven years' programme, and was actively supported by the cannon king, who did all in his power to further his son's interest. An incredible reign of terror began at the Works, where voting papers of a distinctive colour and size were distributed to the men. As there was no secret ballot and the votes were put into the ballot boxes without envelopes, the voting was openly controlled. Krupp officials supervised the voting, under the guise of "assistants to the returning officer," but Stötzel once again proved victorious over Krupp.

After this defeat a policy of sharp reprisals was initiated. In a notice directed against the two local Catholic newspapers, Krupp announced that he considered it necessary to warn his employees against reading these newspapers and to prohibit definitely those who lived in buildings owned by him from doing so. The campaign against illicit readers of the banned periodicals became grotesque. A workman found with one of them in his possession declared that he lived in lodgings and did not notice that his landlady had wrapped his dinner in it; he was, nevertheless, dismissed. Among those overtaken by a similar fate was an old watchman, who had been employed in the works for thirty-three years. The inquisitorial zeal of some of the foremen was so great that they supervised even the toilet paper!

The crowning act of this unparalleled campaign of repression was a demand by the management that every employee should furnish a written declaration as to whether he was a supporter of the National Party or of the Catholic Centre Party. Lists were circulated in the Works that the men's political opinions might be noted against their names. An overwhelming majority of the men refused to comply with the order. In this fashion Krupp vainly tried to maintain the spirit of patriarchal autocracy in an up-to-date industrial establishment.

At the end of the eighties the idea of the "Red Menace" became one of Krupp's obsessions, and when the local authorities suggested that his works were becoming a stronghold of Social Democracy, he fell into an utter state of panic and imagined that his entire business was on the eve of collapse. He insisted on stringent measures being taken by the administrative board in conjunction with the local police to purge the works by dismissing all persons suspected of any Socialist or Communist leanings, without awaiting any actual outbreak. Meyer and Goose became tired of his eternal warnings and interference, and contemplated resigning from the administrative board.

Two attempts on the aged Emperor's life gave Krupp a pretext to demand a fresh purge of the works personnel. His eternal persecution mania gave rise to rumours in the neighbourhood that an attack on Krupp was planned. These rumours induced him to produce another "address," the tone of which was so monstrous that the administrative board declined to issue it. One of the passages from this epistolary masterpiece has been handed down to us: "I will not rest until there is no longer a single Social Democrat in our establishments. . . ."

Krupp remained uneasy till his dying day. Even Bismarck's penal laws against Socialism, causing the imprisonment and exile of thousands of people, appeared too moderate to Krupp's way of thinking. He considered that universal suffrage should be abolished, as the extension of the franchise to all men fostered the growth of Socialism.

Such threats, expressed in the liberal nineteenth century, are singularly like those developed by the rise of Fascism. It must be admitted that the shrewd autocrat fully understood the exercise of power to its utmost potentiality. "There should be flying columns of labour, battalions of young men, available for duty at short notice in mines, factories, or railways," wrote Krupp to one of his directors. His remedy for the outbreak of "the fiendish Lasalle" is remarkably prophetic for 1885 when he says: "I wish some very able man could commence a counter-revolution. . . ."

Fifty years later his wish has been realized.

LAST STRUGGLES

At the beginning of the eighties, when the growth of the American steel industry led to the United States ceasing to be a customer in Europe, the Essen peacetime production of ordinary industrial goods dropped below that of war material. Orders for guns poured in from all quarters of the earth and their value was now counted in scores of millions. Greece, Roumania, Holland, Sweden and Norway, and then Switzerland, completely rearmed their artillery; later came Spain, Austria, Russia, Turkey, Montenegro, China, Bulgaria, Persia, and Italy, the huge guns destined for the last-named state having to be shipped by sea as the Swiss bridges would not stand their weight.

This upward trend of the output of cannon in time of peace made the Krupp works the greatest armament firm in the world. Their total output of guns prior to 1887 amounted to 23,000 and their daily output at that date was 1000 shells, 500 wheels and axles, 450 steel springs, and 1800 rails. The 17,000 employees of 1881 had increased to 20,000 by 1887 and the buildings and plant had been extended in proportion. A shell turnery had been erected and when the steamhammer "Fritz" broke down, 5000-ton hydraulic presses were put in. The new Thomas process made possible the use of pig iron containing phosphorus. The design department now reached its zenith, producing built-up heavy guns, mountain batteries, coast defense cannon, and howitzers of vast dimensions.

The development of the Krupp organization was now at its highest point—but likewise its turning point! Since the Franco-German war Essen had retained its predominant position amongst international arms manufacturers and although never able to do business in England or France, where Armstrong and Schneider reigned supreme, the rest of the world had gradually become a sort of exclusive territory for Krupp. It is just here that a change seemed about to set in. British and French competitors were determined to regain their lost ground and forty years' intensive rivalry amongst great European armament firms now began.

One particular bone of contention was Serbia—the most truculent of the Balkan States. There Krupp was opposed by the French Colonel de Bange, designer for Cail et Compagnie, backed financially by the Comptoir d'Escompte of Paris. Belgrade authorities organized competitive firing tests between Krupp's 3.3 inch, de Bange's 3.1 inch, and Armstrong's 3-inch guns at ranges varying from 1100 to 4000 yards. Krupp was defeated, his gun firing thirty rounds in thirty minutes, whereas de Bange's fired the same number in twenty-three minutes. Furthermore, the breech-block of the Krupp gun was distorted and did not close tightly after firing, which the Essen works described as purely accidental. Further tests failed to remove the unfavourable impression and Serbia ordered the whole of her new armament in France.

This first reverse in the Balkans led to violent recriminations in the press, conducted on both sides with the utmost vindictiveness and very doubtful methods. Both the French and German Governments backed up their armament firms by means of their inspired press. Bismarck's newspaper, the *Norddeutsche Allgemeine Zeitung*, backed Krupp and made revelations regarding serious defects in de Bange's guns, not forgetting to mention the unfortunate victims of the accidents occasioned by them. The *Agence Havas*, the official press organ in Paris, made rather plainer statements: "Colonel de Bange quoted 6.5 million and Krupp 11 million thalers, but when the latter heard of his competitor's quotation, he promptly reduced his to 5 million francs. . . ." In order to enable Krupp to make delivery at a price of 5 million francs and to maintain his position in the armaments world, the German Government were subsidizing him on the contract to the extent of one and a half millions. French industrialists had, however, successfully assailed Krupp's hegemony, as two further contracts from the Roumanian and Mexican Governments had passed by Krupp and had gone to St. Chamond and Creusot, respectively.

Krupp retorted that this statement "is utterly false." He went on to say: "I was never in a position to quote an inclusive price, as I did not know, and still do not know, exactly what the Serbian Government propose to order. . . . As for the two vaunted contracts which constitute such a triumph for French industry, they are actually a couple of experimental guns for Roumania . . . and field guns for Mexico, which I declined to tender for because the commission terms of the intermediary were not in conformity with my business principles. . . ."

The suggestion that commissions and bribes were against his business principles was courageous but not exactly plausible from the generous benefactor of cabinet ministers and generals. In any case he did not need to confine his dealings to small agents, as may be seen from a letter written at about this time concerning a proposed visit to Berlin by the King of Roumania and addressed to his son Fritz Krupp, then in that city: "I shall be glad if the audience is favourable for you, as it is of the utmost importance for our present and future relations with the Ministry. There will be no talk of business, but it might be well to prepare the ground—not for cannon, which we can sell at all times and which will, in any event, be recommended to the King in Berlin, but if you get a chance tell him of our successes with armour plates. . . ."

There is little doubt as to who "recommended" Krupp guns to the Roumanian King in Berlin. Krupp's friends occupied exalted posts in the military hierarchy and civil service and included Bismarck. The effect of their intervention was immediately apparent; in order to make up for the Serbian *débâcle*, further comparative tests were carried out in Bucharest, where Krupp, in conjunction with Gruson, proved "victorious" over his French rivals, and was awarded the contract for arming the forty bastions of the fortress of Bucharest. He achieved similar results in Spezia, where Essen competed with and defeated Armstrong in bidding for the supply of armaments to Italy. These awards were not of actual armament value, but their effect on public opinion was far-reaching. On the surface technical experts investigated and negotiated, while underground, "expenses," commissions, presents, nepotism, and similar honourable devices were the order of the day. Thus in Buenos Aires an article by a Lieut.-Colonel Sellström on the comparative merits of Krupp and Armstrong guns extolled the former—on purely technical grounds, of course—and demanded their adoption for mounting in the new battleship "Almirante Brown." The other side replied through a similarly disinterested technical expert.

In 1886-87 the question of Belgian armaments, including those on the Meuse, loomed large. In this case the local industry of the Liège region entered the competition with an appeal to "national sentiment." The local press declared the Essen system out of date, but here again Krupp found a champion in Captain E. Monthaye of the Belgian General Staff. Wonderful, indeed, are the ways of the armament industry! Captain Monthaye published a monograph on "Krupp et de Bange," in which he endeavoured to prove Essen's superiority. According to the Belgian Captain, de Bange's guns were made from mass-produced Siemens-Martin steel which was inferior to Krupp's crucible-smelted steel in quality and uniformity. The development of a native Belgian arms industry would prove far too costly. The treatise of this unusual G.S.O. concludes with a pæan of praise for Alfred Krupp. His competitors immediately let loose two "disinterested" critics, of whom one, Lieut.-Colonel Hennebert, asserted that the German Emperor and Bismarck were the principal shareholders of Krupp; while the other, Lieutenant Malengreau, accused Krupp of producing crucible-smelted steel only when observers, more especially newspaper reporters, were present. The authorities in Brussels steered a middle course through the fire of this mercenary hysteria; they ordered heavy ordnance from Krupp and called for tenders for the light guns, these being awarded to his competitors.

Such peaceful solutions were by no means the rule,

as the great arms-manufacturers' rivalry aimed at something more than a profit in their accounts; they sought to make the contracting state dependent on them for all future supplies of arms. This necessitated the support of their own governments, and it was frequently difficult to determine whether the latter's policy was inspired by business negotiations or whether these were dependent on the government policy. This was particularly so in the case of Krupp's growing trade with Turkey, where Essen, backed by Berlin, was always ahead of its western competitors. Hallgarten writes on this subject:—

"Germany's interests in Turkey, which, as the oldest opponent of Russia, is naturally inclined towards any political rival of that country, are inspired by definitely capitalistic motives. The mission of the Prussian General von der Goltz, sent by the German Emperor at the request of the Turks to reorganize their army, had also the further task of forcing Krupp cannon on Turkey."

Alfred Krupp looked upon the German Ambassador in Constantinople as an unofficial representative of the Essen firm. When the Sultan and the Grand Vizier were to be presented with an album containing illustrations of Krupp guns, he wrote to his board in Essen: "Undoubtedly the Ambassador, who can easily discover our relations with the Emperor even if he does not already know of them, will give any necessary advice, indicate ways and means or act as intermediary himself."

In July, 1885, Turkey ordered some 500 guns and placed further contracts in the following February for the supply of 426 field guns and howitzers. Constantinople Customs House revenues for several years were assigned for the purpose of defraying the cost, amounting to many millions, of these purchases. The object of

Turkey's intensified armament policy was revealed in the contemporary revolt in Philippopolis, in the annexation of Eastern Rumelia by Bulgaria and the consequent grave European crisis. Sofia was supported by Russia (the banking firm of Günzburg and the railway contractor Poleskoff had interests there) while German Imperialism was paving the way for the completion of the projected railway from Berlin and Vienna to Philippopolis and Constantinople on the Bosphorus, where Krupp guns barred the passage of the Russian Fleet.

The French armament industry, which, despite its success in Serbia, had met with a series of reverses in those years, now proceeded to a successful assault on Krupp's position in Japan. This was facilitated by the close friendship between China and the Essen firm, and the final blow was administered by the "Naval Adviser" Bertin, who was dispatched to Tokyo by the French Foreign Office ostensibly as such, but actually as a business representative of the firm of Schneider. Bertin took good care that all Japanese ships built after about 1880 should—despite Krupp's successful gunnery displays carried out before the Mikado—mount practically nothing but guns made in Le Creusot. In conjunction with the penetration of French capital into Russia's industries this success of Schneider is the first serious blow against the Krupp world monopoly.

Having embarked in an armament conflict with Russia and France, there remained one power still untouched during the closing era of Bismarck, *i.e.*, Britain. In this case the Essen sales representatives embarked on an intensive campaign of direct action as opposed to the petty pinpricks used on the Continent. In 1879 Krupp circularized the members of the House of Commons "to correct erroneous impressions" regarding the failure of his system of construction. When naval guns burst, competitors on either side of the Channel abused one another's work, although they were busy copying each other's designs. Colonel Maitland, Director of the Royal Gun Factory in Woolwich, himself wrote (in the Journal of the Royal United Service Institution) how, in the course of a visit to Essen, he measured parts of the 15.75-inch gun which he saw in process of construction, with his umbrella. The relatively friendly nature of the rivalry between the armament firms is illustrated by the fact that a son of the house of Vickers served part of his apprenticeship in the Essen Works. But the friendly atmosphere underwent a radical change the moment Krupp attempted to attack British spheres of influence outside Europe.

First came China, where the Essen firm secured an important contract for rails in the face of British competition. The Chinese Premier (and future Viceroy) Li Hung Chang severed his connection with the British armament industry and openly favoured Essen. Alfred Krupp wrote him flattering letters and presented him with a beautiful working model of a railway. China became such an important customer that Krupp suggested asking the Foreign Ministry whether letters should be sent direct or forwarded through the German Legation in China.

The resulting business was a trifle disappointing for Essen. China purchased *only* 150 fortress guns and 275 field guns, in addition to the complete armament of eight battleships! But future prospects were rosy and Li Hung Chang's portrait at Hügel occupied the place of honour —over the bed of the old gentleman! He clearly appreciated the fact that Li hated the British, but not the Germans whom he wished to use against them. London was also alive to this situation.

Essen's home trade was greatly assisted by the new tariffs policy. Orders for rails and wheel tires kept coming in in growing numbers and they showed a huge margin of profit, as the elimination of foreign competition enabled home producers to raise their prices by over fifty per cent. The reverse side of the picture was the fantastic dumping campaign in foreign countries. As regards rails, the existence of the International Rail Cartel presented an obstacle until 1886. Then it expired and was not renewed. Krupp, who always claimed that he maintained his prices, now became prominent as a dumper. This is revealed by the current report of the Essen Chamber of Commerce, which states: "The Works situated in our district have secured important contracts in the international market in the face of competition by foreign manufacturers. At best the selling prices equal production costs, but in the majority of cases—more especially where mass-production articles are concerned—the prices realized are below manufacturing cost." Dumping was particularly indulged in with rails for the Italian Mediterranean Railway, where Krupp, in association with Franco-Belgian firms, underbid the British.

At the great expositions of those years in Sydney, Melbourne, Amsterdam, Berlin, and Düsseldorf, the Essen Works again appeared in the guise of a rising industrial giant of the highest class. How far off seemed crises and reverses! The large loan contracted some time previously was completely repaid by the firm between 1883 and 1886—long before its repayment was actually due. This represents a sum of twenty-two millions in three years paid out of profits, and during the same period Krupp extended his activities by purchasing the steelworks of Asthower and Co. in Annen in November, 1886. These works specialized in steel diecastings for machinery parts and components of rifles and revolvers of a highly intricate nature. The causes which compelled the successful engineer Asthöwer to bend his neck under the yoke of his Essen competitor have never been disclosed. In the course of the same year Krupp concluded an agreement with the Lorenz Metal Cartridge Case Works in Karlsruhe, under which he secured the selling rights of the metal cartridge cases which formed such an essential accessory for his guns. In 1896 the Lorenz works were amalgamated with the notorious Deutsche Waffen- und Munitionsfabriken A.-G., to which the Essen works were therefore bound by close ties. Then the first contracts were made with the International Powder Trust. When the famous Nobel had invented his smokeless powder, he came to Essen in person to have the first bags tested by Alfred Krupp. The latter was delighted with the result and acquired the rights for Germany.

Despite this phenomenal expansion, the Essen horizon was by no means cloudless. Their business methods no longer passed unchallenged and public opinion became critical of the monopoly Krupp had created for his firm. One sequel to the hostile sentiments it evoked was a violent conflict with the Admiralty, which marked the end of the old man's life. Its origin was a real menace to Essen. A Krupp built-up gun on board the training-ship *Renown* exploded and an inquiry into the cause of the accident led to a thorough examination of the remaining cannon, several of which showed serious fractures. The Admiralty became alarmed and opponents of cast steel guns once more raised their voices, and pressed the Admiralty to insist on Krupp guaranteeing his guns for a minimum life of 500 rounds each. This frightened the cannon king, who was fully aware of the difference between making extravagant claims for the excellence of one's wares and giving a proper legal guarantee in respect to them. He declined to furnish such guarantee and entered into an increasingly recriminative correspondence with the Chief of the Admiralty, von Stosch.

The old man showered daily admonitory letters on his own people. To his dismay he found that his blunt refusal had no supporters; the administrative board and the Berlin representative were in favour of meeting the Admiralty half-way as they could not afford to quarrel with so important a customer. Irritated by Krupp's truculence the Admiralty now announced that if the provocative attitude of the Essen Works was based on their monopoly the naval authorities would take steps to correct their error in that regard.

This sounded threatening, but failed to impress the stubborn old man, who dispatched Jencke, the new chairman of the Administrative Board, to Berlin to make confidential statements to the Emperor and to Bismarck. In these statements Krupp announced that he would retire from business and sell the works unless the Admiralty gave way. The effect of his trump card was immediate—under pressure from the Government the Admiralty waived their demand for guarantees, provided improved forging and more thorough supervision of manufacture were promised. This opposition was the last flare-up of Old Prussian resentment and it ushered in a new era of harmonious co-operation between business interests and Government bureaucracy.

Where the struggle was for power and profit, the cannon king retained his uncanny shrewdness up to the end, but he had become a definite obstacle to technical progress in the works. This obstructiveness on his part was nothing new. Alfred Krupp, despite his early training in his father's works, was never a technical expert. His innumerable letters contain ample evidence of his brilliant business ability, his perspicacity in regard to financial matters and knowledge of human nature, and there are, of course, plenty of observations relating to technical matters. But the attempts made by Alfred Krupp's biographers to laud him as an engineer of genius—both as regards his early work in connection with metallurgical research and his later activities as a mechanical designer—are singularly unconvincing. All these writers are able to prove is Krupp's ability to exploit other people's ideas (*e.g.*, the spoon roller and railway wheel tire) together with his utter lack of technical initiative for improved methods of production or uses of new materials. Krupp himself never developed a really great invention connected with cast steel or with gun construction.

He appears to have admitted as much himself at times, if his derogatory remarks concerning the financial abilities of inventors are any criterion; in fact he openly said: "Let us leave inventing to other people; it is better and less expensive to utilize the completed inventions of others and pay a reasonable price for them."

As an example of his lack of technical knowledge the case of the armoured gun may be cited. Alfred Krupp bickered with his works staff for twelve years about this gun, which was constructed on a very "simple" principle —instead of projecting through a slot in the armour-plate, the gun muzzle was mounted on a movable ball joint in the armour. In practice this arrangement suffers from the disadvantage of producing a very limited arc of fire, and however good the protection afforded by the heavy armour may be, it is a nuisance in an attack. The material crushes the construction. Notwithstanding the scepticism of the gunmounting shop, the Works proprietor devoted an immense amount of time and money to his idea. Over 4000 sheets of sketches and notes on it were found among his papers. Proud of his creative talents, he told the Emperor about it and, disgusted at the apathy of the technical experts, drafted a provocative letter to the Admiralty which his administrative board implored him not to send in its existing form. His senile stubbornness prevented him from seeking advice from Gross—the one man who might have been able to help him. The whole proposition proved to be technically impracticable and no orders worth mentioning were ever received for armoured guns.

The same thing occurred in regard to another idea evolved about that time—the armoured boat. The main idea was equally startling: a huge gun was to be mounted on a tiny boat. The question as to how this was to be achieved was for the naval architects to solve. The Minister of Marine, von Stosch, vainly attempted to persuade Krupp that the probable effect of the recoil on the relatively flimsy hull would be to damage or even totally destroy the latter, and practical experiments confirmed these misgivings. But Alfred Krupp did not abandon his pet schemes so easily. Once before he had designed a "funnel" to divert the blast of the armoured gun upwards and he now proceeded to develop an idea which even one of his most admiring biographers describes as "crazy." Krupp proposed to mount a single gun in the armoured boat—which could fire in two directions simultaneously! One shot was to be fired at the enemy and the other into the water. The latter was of course superfluous, but was to absorb the recoil. Engineers were at pains to explain the absurdity of such a scheme to him. "You probably think I am crazy, but if you imagine that I care anything for such opinions you are mistaken," he replied. Finally this scheme, like many others, petered out.

Once more the old man's restless imagination conceived the idea of using heavy guns, unprotected by armour, at sea. Failing everything else, he evolved a "floating battery," a raft mounting guns and rendered unsinkable by means of airtight compartments. As such a monstrosity would not be equipped with any propelling machinery, it would have to be towed or pushed along by real ships. This fantastic idea kept a number of engineers busy at the old gentleman's personal disposal for quite a long time, but it was finally buried with him. And the one innovation to naval armament which he declined to entertain, the ironclad ("a fantastic and expensive device"), was destined to have a great future.

Further light on Krupp's qualifications as an engineer is shed by his behaviour to Gruson. Next to Bochum die-cast steel, Gruson's hardened steel constituted one of the most important metallurgical developments of the day. But Krupp had always regarded it with derision and because of it, had acquired the dangerous habit of under-estimating the importance of further successful discoveries of the Madgeburg firm. When Meyer called his attention to the new Gruson armoured turret, in which the usually reserved Prussian military authorities were displaying increasing interest, he put forward a number of silly objections. He talked of "an old pot," although the flat top of the turret was a masterpiece of design in contrast to his own armoured gun. "Why should we worry about the majority of blockheads!" is how Alfred Krupp swept aside the considered opinions of specialist engineers. Armour-plate and shot made of hardened steel now began to compete successfully with the products of Essen. Gruson, in conjunction with the brothers Siemens, planned the building of a great steel-works in Saxony where cannon also could be made. The scheme failed to materialize, but the competitive production of hardened steel shells caused Essen quite a lot of worry. In desperation Krupp abandoned his standard specifications and insisted on entering into competition with Magdeburg by "casting at rock-bottom cost," which the works declined to do. He also endeavoured to develop his grotesque "bar armour," in a vain attempt to catch up with Gruson.

Krupp's antipathy to the man from Magdeburg became a mania, and when he was seeking an engineer (he wanted "the best qualified one in all Prussia") he stipulated that he would not consider any candidates who believed in Gruson. He met Major Schumann, a distinguished expert in the construction of armour and of fortifications, and the latter put forward a suggestion, which was backed by Gross, that turrets of Gruson pattern should be built. But the stubborn old man would have none of it. He ignored the representations of both brilliant experts and the permanent employment of Schumann became impossible. The disgruntled expert went to Magdeburg where his ability was recognized. Much of Gruson's success with his later designs was due to Schumann.

Old Krupp was still eager to deal what he hoped would be a deadly blow at his rival. He proposed to carry out public comparative tests between his own steel products and the inferior ones of Gruson. The following account of these tests is given by the pro-Krupp biographer Frobenius:—

"He did not consider it necessary to obtain a plate from Gruson, as he declared that anybody could manufacture similar hardened steel plates. He had one made in his own works—which was a great mistake.

"He then set up a section of his own bar armour-plate and a segment of a hardened steel turret, the former being 20″ thick, backed by an armoured grating 24″ thick of 36″ diameter, while the latter was only 25″ thick. As he fired at both with the same 10″ gun, he should have taken a much weaker plate of his own. . . .

"The arrangement proved Krupp's complete ignorance in the field of armour-shooting tests . . . for whatever the results, they were certainly open to objection."

It would appear, therefore, that Krupp's tests were carried out with faked Gruson plates, which allowed him to demonstrate the superiority of his own armour. One cannot help wondering what he would have said if somebody else had manufactured "Krupp guns" in order to demonstrate their inferiority. But even such questionable methods of competition proved useless against the Magdeburg man's great invention and towards the end of his life Krupp was compelled to see the despised hardened steel plate generally adopted, while Essen's would-be competitive product had proved a hopeless fiasco.

Alfred Krupp was unable to appreciate the growing importance of armour-plate in modern armaments. While three separate armour-plate systems—Sheffield's compound plates, Creusot's steel plates, and Gruson's hardened cast steel plates—competed against each other, he held aloof. He would produce the best guns and the hardest shot, but wished to leave armour-plate to others. The engineers at his works, who did not share this prejudice, were waiting only for the old man's death to remove the obstacle to their plans.

LONELY DEATH

For many years past Alfred Krupp, known to the world as a genial, energetic dictator, was in actual fact a misanthropic old man, who seldom set foot in the works and frequently remained out of touch with their administrative branch for weeks on end. He communicated even with his wife in writing.

His marriage had not been a happy one and the lack of common interests had, after his wife's prolonged illness following the birth of their only child, caused them to drift apart. The wife was fond of society and of entertaining, while the churlish Alfred Krupp disliked any kind of relaxation or social intercourse. Mrs. Krupp's hopes of a new form of life consequent on their move into Hügel Castle were not realized and she absented herself more and more from a place and community which she both disliked and despised.

Occasional trips together and affectionate letters,

mostly concerned with the care of their only son, failed to disguise the growing estrangement between husband and wife.

Krupp had worried about his health from his earliest days and every letter he wrote while on his travels contained references to some bodily ailment or other. As he grew older, this hypochondria became more and more marked and he frequently took to his bed for weeks at a time, although his medical advisers were unable to diagnose anything wrong with him.

The everlasting sickroom atmosphere, together with the querulousness of the old valetudinarian, disgusted the young wife to such an extent that she finally left him entirely. The sole contact she maintained with her husband was through their only son.

In his later years old Alfred Krupp became extremely eccentric and his tiresome inhibitions limited the circle of his visiting acquaintances to nonentities and self-seekers. He did indeed try to find some distraction in seeking the acquaintance of artistic people in Düsseldorf—including that of the young actress Franziska Elmenreich and the celebrated pianist Franz Liszt, one of whose pupils, an attractive young woman, was assisted by Krupp. When sinister rumours concerning the motives for his patronage of this lady began to circulate, Krupp—to whom Liszt himself had mentioned them—protested vehemently. Liszt quietly replied that he did not lay claim to any right to judge the morals of his pupils.

The seventieth birthday of the cannon king found him completely isolated. His brother Hermann, whose friendship he repeatedly rejected, and his sister Ida were both dead. Krupp's friendlessness was not only due to deaths. Communication with his surviving brother was maintained through the works management. His old coadjutors Meyer and Goose had retired, disgusted with his stubbornness. His relations with old Emperor William in Baden-Baden had become less cordial. Krupp maintained that the Emperor had been prejudiced against him, but it is just possible that the shamelessly mercenary character of the cannon king had, at long last, revealed itself to William. Perhaps Krupp no longer valued his friendship. In a letter to his son he wrote: "You must be on the same footing with the future Emperor as I used to be with the present one. . . ."

Alfred Krupp was gradually losing strength and spent most of his time in bed. The old man was very tired, he even gave up the habit, indulged in for many years, of writing daily letters and memoranda.

He was, however, still disturbed about business matters. Gross's prominent position worried him, as the brilliant engineer did not conform to his ideal of an obliging employee. He would have liked to subordinate him to a "technical administrator," but the Austrian Kuppelwieser declined the appointment and he himself had misgivings about Adolf Kirdorf, "an officer of a public limited company—which is equivalent to his being a swindler!" The only man he really trusted was Longsdon, whom he had always held in great esteem. He commended the Englishman to his son, saying, "We did not get a real start until we introduced British methods." Contemporary schemes for constructing canals alarmed him; he wanted to sell rails and to keep down the price of coal by localized distribution, so canals could only be a national calamity.

At Easter in 1887 he dragged himself down to the works for the last time. He noticed that the workmen living on the Kronenberg Housing Estate had built goat-sheds against their houses. Where could they get their fodder? They stole it, of course. Krupp's opinions of "the lower classes" had grown worse: "Nowadays the people are happy only when they have nothing to care for. The women put everything on their backs, the dairymaid apes the lady, and the men spend their money on liquor. . . ."

In July Fritz Krupp came from Heidelberg to see his father, but misled by the latter's apparently robust appearance, he went away again.

The following day, July 14th, 1887, Alfred Krupp died from heart failure, at the age of seventy-five. He died in the arms of his valet Ludger.

Torches flared in the midnight darkness as the body was carried to the factory, to lie there in state. Even his body was to bear witness for the "ancestral home," this symbol of small beginnings, in a period of dangerous proletarisation. But even that is deception, for the dead man sprang from an old family which had always enjoyed power and place. He was the climax of their will to rise.

He had long since seemed like some obelisk of a past age. Born when his father pledged allegiance to Emperor Napoleon, then about to embark on his disastrous Russian campaign, he took over the direction of the works at a time when Germany was an impoverished agricultural country, with the steam engine and the railway still considered mere engineering fairy tales. When he became owner, the first wave of civic revolution swept through reactionary Prussia; he began manufacturing arms when William and Bismarck were inaugurating their *régime* of blood and iron. He had lived a score of years under the reconstituted Empire, piling up power and wealth. He had never been able to understand modern industrialism in which science governs production, its social problems, strikes and class warfare. Nor had he understood the Empire which, looking beyond national self-sufficiency, strove for world supremacy beyond its frontiers.

The Cannon King—a title his son also inherited—represented the first great era of German industrialism. Those who went before, Dinnendahl and Harkort, were the prophets and pioneers of the Age of the Machine. While those who followed, Thyssen, Stumm, Kirdorf and Stinnes, were the reapers and the organizers of a top-heavy capitalistic trust. Alfred Krupp stands midway—a conqueror and mighty robber, endowed with all the energy and ruthlessness for which his task called. Politics, philosophy, and the arts were as foreign to his mind as were the slightest doubts concerning the divinity and eternity of the social order, to the building of which he himself had contributed so much.

Legends were woven even before the grave had closed over him. Jencke, as Chairman of the Administrative Board, referred to "the dear, kindly and noble gentleman, a shining example of that ardent patriotism which feels no sacrifice too great for his country." Thousands of newspapers and periodicals wrote in a similar strain. In the *International Review of All Armies and Navies* it was stated: "Krupp's patriotism would not permit him to furnish effective weapons to France for use against his own country." Those were fairy tales—but they were believed! A few moments' study of the fifty-three high decorations awarded to the Cannon King might lead to different conclusions. Besides the Prussian Orders of the Eagle there were the red ribbon of the French Legion of Honour, the Grand Cross of the Spanish Military Order of Merit, the Russian Order of St. Anne with Diamonds, the Grand Cross of the Belgian Order of Leopold, the Corona d'Italia, the Japanese Order of the Rising Sun, the Turkish Medjidie Order, and the Brazilian Grande Dignitario. Hügel Castle contained many treasured keepsakes, including a diamond ring from the Grand Duke Michael, a gold snuff-box from Francis Joseph of Austria and the 2000-year-old vase from Li Hung Chang—all tokens of recognition of services rendered, of gratitude for arms supplied, and of technical advice on their construction. Of the 24,576 cannon made in the works up to the time of his death, only 10,666 were destined for Germany, the remaining 13,910 being sent abroad. Many of them were to be turned against their maker's country in that fateful August of 1914.

Krupp's will directed that the entire works should be entailed in trust for his heir and the latter's descendants, thereby ensuring that sole possession should be vested in them for all time.

Alfred Krupp was no great inventor, no revolutionary in technical processes. His chief quality, not always understood by his most ardent biographers, was that he was a gifted merchant, that he understood how to sell his wares, and how to change production with the changing needs of the times. And he understood as well that it was not enough to be just one more profit-hunter, to sell *his* sort of wares without an "ethical" background. He surrounded his Works with the glamour of philanthropy and made them . . . which took courage . . . a national institution.

PART IV

THE HEIR

Armour Plate and Cannon—The Navy Era—Krupp, Member of the Reichstag—Concerning the Recoil Cylinder—The "Leap in the Dark"—Capri

THE HEIR

ARMOUR PLATE AND CANNON

THE very name of the man who was now master of Germany's greatest arms factory was characteristic. As a boy he was called "Fritz," and at the age of twenty he received letters addressed to "Fritz Alfred." Finally he became known as "Friedrich Alfred," upholding memories of the founder of the Works and the creator of their greatness. The heir had been thoroughly trained for the part he was to play before the eyes of the world regardless of his inability to bear the burden it entailed.

Fritz Krupp was the son of an elderly father and of a mother whose health was permanently injured by his birth. Born February 17th, 1854, he was delicate in childhood and suffered from asthma. He grew up a shy and highly sensitive youth who had little in common with companions of his own age. His education was subjected to frequent interruptions through illness.

The Cannon King worried much about his son's ability to take over the direction of the Works. He even considered bringing one of his most talented nephews over from Berndorf, although his innate distrust of his brother finally caused him to renounce this idea. Long journeys were undertaken to consult physicians about the boy and to obtain treatment for him. Such was his youth.

The gloomy days in Torquay and the shock of the depression brought father and son nearer to one another.

Fritz began to keep a diary in which he entered advice given him by the old man. The latter was jealous of anybody who might possibly supplant him in his son's mind and he emphatically vetoed Fritz's wish to study natural science, for which he had a real gift. Old Krupp did not believe in scientific training, he left that to his paid employees; he was more concerned with training his son in the administrative management of the Works and constantly admonished him to acquire a thorough grip of the whole business, more especially in regard to questions of personnel and matters of finance. He viewed Fritz's entry into the Karlsruhe Dragoon Regiment with great disfavour, but refrained from taking any action as he was unwilling to lay himself open to the charge of having his son excused from military service. However, after a few weeks' service, Fritz was discharged from the army as medically unfit—the Cannon King's son would never be exposed to any risk of death or injury from cannon balls.

Barely twenty years of age, Fritz was entrusted with his first independent job—chiefly as representative of the firm. In the course of a visit to Egypt he made offers to the Khedive for the construction of a great railway to the Sudan. He represented his father on the occasion of the Emperor's visit to the Düsseldorf Exhibition and attended him at Hügel. Later he called on the bloodthirsty Sultan Abdul Hamid, presenting him with a flattering address which praised him as the benefactor of the Turkish people.

At the Works he gradually became intermediary between the progressive administrative board and his stubborn, reactionary father. The old man was softened a bit by the sudden departure of his wife. He allowed

Fritz to attend a short course at the Technical College in Brunswick and appointed him to the administrative board. In 1882 he agreed to his marriage, which he had long opposed. Young Mrs. Margarethe Krupp was a daughter of Baron von Ende, Lord Lieutenant of Düsseldorf and a neighbour of Hügel. She is said to have been able and energetic. The young couple took up their abode in "The Small House" of Hügel, where Bertha Krupp, present owner of the works, was born in 1886 and also a second daughter, Barbara.

The death of Alfred Krupp, followed twelve months later by that of his wife (which passed unnoticed), brought drastic change of policy at the Works. Although the administrative board directed the management and exercised full executive functions in control of the complicated organization, the word of the owner had always had the force of final law. The Cannon King grew up under primitive conditions in industry, he was always narrow-minded, prejudiced, and latterly opposed to any innovation. Young Krupp, accustomed to different surroundings, had a broader outlook and a more accommodating character far easier to deal with. All the forces hitherto repressed by the old man were now allowed full play and improvements and modern methods adopted in all directions. Fritz Krupp was fully aware of their necessity if the Works were to hold their own with younger competitors. The administrative board, to which the ordnance designer Gross was at last admitted, was given a free hand. It was invested with full powers in all matters appertaining to representation, production, and business in the widest possible sense.

The undertakings of the Works during this period showed speed and scale. The number of Martin furnaces was increased, the smelting shop enlarged. A fifth gun-mounting shop and enlarged artillery halls indicated increase in the production of armaments, while a third axle turnery was necessitated by many new orders for railway material. A number of collieries situated in Lorraine were acquired to meet the increased consumption of coal.

In embarking on the manufacture of armour-plate for land defence purposes, Krupp came up against a rival who had hitherto enjoyed a virtual monopoly of this particular branch of the armament industry—the Gruson works. Old Krupp completely failed to combat them. While he derided hardened cast steel as "cast iron," the man from Magdeburg developed it with success. He produced shells with increased penetration at a low cost; and while Alfred Krupp wasted time over his clumsy armoured gun (essentially a problem of materials) Gruson turned out his first armoured turret, a real masterpiece of engineering. The Cannon King unwittingly contributed to his rival's great achievement by his brusque treatment of Major Schumann, who then went to Magdeburg. Gruson was shrewd enough to give the stubborn inventor a free hand. He took over his designs for fortress armour, the flat-pressed turret in which a gun could only be moved vertically, but the turret could be rotated and easily sunk. Armoured turrets built to Gruson-Schumann designs have been universally adopted for use in modern fortifications.

The conflict between the two competitors was sharpened when Gruson began to manufacture automatic guns, quick-firing howitzers and field guns, and equipped his own range at Tangerhütte. Essen knew that a competitor was out to challenge its position in the German armament industry.

Cautious plans were laid to trap the presumptuous rival. As already related, the comparative firing tests carried out in 1879 in which the Cannon King used faked Gruson armour plates to engender distrust of Magdeburg, illustrated the lengths to which he would go to weaken a competitor. After such preparation came the main attack. It was considered the proper thing in Essen to express indignation over "Stock Exchange swindles," but that did not hold good for conflict with competition. By the help of cleverly disguised dummies, Essen proceeded to buy up every Magdeburg share that came on the market. In order to keep the price down, the utmost caution was used in the transactions, until Karl Fürstenberg of the Berliner Handelsgesellschaft and the Diskontobank were in a position to report that the scheme had been successfully completed. At the general meeting in the spring of 1892 of shareholders of the Gruson A.-G., the astonishing fact was disclosed that Krupp held a majority of the shares. The might of money had defeated technical superiority.

The general meeting, now dominated by Krupp, first authorized a working agreement, which led to an outright sale twelve months later and in May, 1893, the Magdeburg works were incorporated in the Essen concern under the title of the Fried. Krupp-Gruson Works. This callous procedure gave rise to much newspaper comment, in which the cynical duping of shareholders who were induced to part with their holdings at thirty per cent. below their proper value was mercilessly denounced. Hermann Gruson viewed the collapse of his life's work with bitter resignation and he never again set foot in the works he had created.

About the same time the first armour-plate rolling mill was opened at Essen. The vast modern steel and iron building in which it was housed completely dwarfed the older low-built shops which surrounded it. It contained gigantic machine tools, huge overhead cranes, and hydraulic presses of 10,000-ton capacity. A few years later the erection of a second armour-plate shop became necessary, as Essen traditions were opposed to waiting for business before providing the means of dealing with it. One must always be ready to attract business by suggesting new methods of production. For the next decade business was ingeniously stimulated by a cleverly staged competition between armour and guns in which Krupp appeared as the creator of his constructional designs.

From the beginning of the nineties he made guns from a nickel-steel alloy perfected by the American Harvey. It proved to be exceptionally tough and supplanted the earlier brittle varieties of steel. When rolled into plates, it maintained its quality and showed neither cracks nor fissures when pierced by shells. This relative softness had certain disadvantages, as it is the function of armour to withstand shells and not to let them through. The problem was to combine the properties of internal toughness with a hard outer surface and Krupp succeeded in solving it by means of "oil hardening" which increased the resistance of nickel-steel plate by ten per cent.

It is obviously not in the interests of an arms firm to evolve a perfect armour which would make guns and shells useless against it. The aim of the armament firm is not the solution of technical problems but the making of profits. Business can flourish only if both sides in the artillery duel, attack and defence, are equally balanced in the eternal struggle for supremacy. Krupp displayed uncanny skill in alternating the lead between armour and gun. He opposed the oil-hardened nickel-steel armour-plate by projectiles made of chrome steel, so hardened by a special process as to enable them to destroy the toughest nickel-steel plate ever made.

Then it was again the turn of armour-plate. The problem was not an easy one, for though steel of high carbon content could withstand the new shells, it was too brittle to be relied on not to splinter or crack. Krupp therefore tried to "enrich" the carbon content of the surface of soft nickel-steel plate in order partially to "harden it." Armour plate of this kind was exhibited at the World's Fair in Chicago. Further improvements were effected by hardening with water and subjecting to high pressure, until a surface hardness like glass was obtained. At the great experimental test firings of the nineties the firm demonstrated this ostensibly perfect armour plate and offered it to the world as "Krupp-Armour" with a thirty per cent. superiority over Harvey steel.

This last word in the field of armour did not remain unanswered by gun specialists. It is particularly interesting to note that the firm carefully timed the production of armour and shells to produce the greatest effect. In matters of actual ordnance construction Krupp was definitely lacking in originality, as was soon to be proved. But, on the other hand, his factory produced improved projectiles. These were "capped shot" having an explosive nose, which was intended to burst inside the armour and utterly destroy the latter. Again the supremacy of the attacking gun over the defending armour was re-established.

Each of these technical developments produced a huge volume of orders and enormous profits for Krupp. The Essen concern had, unnoticed by its contemporaries, become a sort of barometer, which indicated variations of atmospheric pressure in the international armament race —in fact it frequently foretold them. This is most clearly illustrated by the great expansion of Krupp's undertakings through the purchase of the Germania Ship Yard in Kiel. This concern, originally Danish, had passed through many vicissitudes and frequently changed hands. Essen's sudden interest in the shipyard was not unconnected with the naval policy of the Government. Exactly two years before the first great German Navy Act, the start of a new era of German naval construction, the Essen works acquired a lease of the shipyard which was later completely absorbed under the title "Fried. Krupp-Germania Ship Yard." The uninformed shareholders were thoroughly swindled by the well-informed House of Krupp. Presumably acting on a hint of favours to come, Essen spent millions in reconditioning the yard. Influence in the right quarters caused the usually bureaucratic Admiralty to display a singularly accommodating spirit in the cession of adjacent land, giving the shipyard a half-mile of waterfront. The place was soon converted into a thoroughly up-to-date dockyard capable of building and equipping a warship with armour, guns, boilers, engines, and all accessories. Orders would not be slow in coming, either.

The number of Krupp employees, amounting to roughly 20,000 in 1887, increased steadily at the rate of a thousand a year. This immense growth was not peculiar to Essen. After the great crisis German industry emerged slowly from a lengthy period of intense depression. A minor crisis in 1887 gave it a further setback, but following the accession of the last Emperor a period of rapid industrial expansion commenced such as the country had not seen for many a year. Bismarck's successor, General Caprivi, began by renewing the various German commercial treaties on terms which were intended to safeguard the interests of the great industrialists. During the fifteen years ending in 1902 protective tariffs increased German production of iron from four to ten million and that of steel from one to eight million tons.

This increased output was not due to any Balkan conflict or North American railway construction boom; on the contrary, whereas Krupp's exports between 1875 and 1890 comprised nearly two-thirds of his output, his exports were now smaller than his home sales. As may be easily guessed, this extraordinary change was largely due to the growing spirit of militarism and to naval construction programmes involving an expenditure of many millions. Germany's era of imperialism had begun.

THE NAVY ERA

The dismissal of Bismarck removed the last check on the Government's policy of world expansion. The party favouring colonial expansion and the creation of a huge navy had the ear of the Emperor. Their views were shared by the still unknown Tirpitz, the coming man struggling for recognition behind the scene, who, in a memorandum written in the autumn of 1891, clearly expressed the view that a mighty battle fleet was essential for Germany's future.

The "Left" parties in Parliament offered opposition to this naval programme and throughout the next decade maintained a struggle which caused the Admiralty to delete many a ship from its proposed estimates. Although the admirals obtained approval for the construction of some two dozen armoured cruisers, they were not satisfied with the progress of Germany's naval armaments and bombarded Government and Parliamentary parties with memoranda in which they attempted to prove a decrease in the strength of the German navy. The climax was their demand for an extended programme of naval construction at a cost of 150 millions and the opposition to this demand was the last serious effort ever made by the Liberals of Germany. The member for Essen, Mr. Hammacher, announced that if Parliament rejected this navy programme there would be a *coup d'état* on the part of the Emperor.

However, no open conflict, similar to the one over the constitutional issue raised by the Army reforms, arose, as the Parliamentary struggle was unexpectedly taken up elsewhere. In March, 1897, a cable was sent to the China Station appointing Rear-Admiral Tirpitz, commanding the cruiser squadron there, Secretary of the Admiralty. The keenest and ablest supporter of the "big navy" programme was given a responsible post, and, surprisingly, he utilized it by beginning a widespread campaign for popularizing the navy. The industrialists concurred in his scheme and with Krupp at their head they proceeded to create a new medium for propaganda —the German Navy League.

A worthy middle-class nationally minded citizen, the Berlin cod liver oil merchant J. E. Stroschein, had, in an access of disinterested zeal, begun to organize a Naval

Association, but before he had succeeded in fully realizing his idea it received a counter-blast from a wholly unexpected quarter—from Essen. The administration of the Krupp works were, so says the naval historian Eckart Kehr, of opinion that "this Association would develop along ideologic lines, which would take insufficient account of the interests of industry." Honest Stroschein seemed unlikely to lend himself to the schemings of contract-hunting naval enthusiasts, so away with him! Under the auspices of Essen a rival organization, the "German Navy League," was called into being, with ample funds for exploiting the original aims of the Stroschein Association while discarding its idealistic founder.

The president and society figure-head of the League, which was organized on April 30th, 1898, was Prince William of Wied (the future King of Albania), but its real chief was the general secretary Victor Schweinburg. This Jewish journalist from Moravia could boast of possessing the best possible contacts with the Essen administrators. He edited the *Berliner Neueste Nachrichten*, acquired by Krupp in the early nineties in order to have a mouthpiece in the capital of the Empire. The unscrupulous manner in which Schweinburg contrived to convert the business interests of the firm into affairs of patriotism fully qualified him for the task of disguising the newly established publicity office as a popular national movement.

Money flowed in freely. Its source was disclosed by the conservative *Kreuzzeitung;* the subscribers were not the honest and plain folk, but "mostly industrialists and business men who wished to make money out of the naval expansion policy." Needless to say, Krupp was the foremost of these. When it was revealed at an annual meeting of the League that "private persons in Westphalia have made donations amounting to half a million" a motion was put forward to decline such contributions from people who were financially interested in the reinforcement of the Navy. The motion was, of course, indignantly rejected. In order to conceal the extent of this corruption the accounts of the League were said to have been destroyed later. But, like Charles M. Schwab and John Pierpont Morgan, the men behind the American Navy League, Krupp was ever shown to be fully identified with the combination of business interests and naval propaganda. His frank admission of this fact is indignantly related by Heinrich Rippler, the editor-in-chief of the nationalistic *Tägliche Rundschau*—

"Mr. Krupp presented Mr. Schweinburg to the Emperor and the photographers have recorded this momentous incident for posterity." Again, a little later: "At a meeting attended, among others, by State Secretary Tirpitz, Professor Schmoller declared, amid the thunderous applause of the entire audience, that he had nothing in common with Mr. Schweinburg—but the leader of the ironmasters, Mr. Jencke (head of the Krupp administrative board) took this man to the Emperor and praised his press agent so highly that His Majesty showed his gratitude by shaking Schweinburg's hand."

The only important effect of this statement was a protest by the professors of the University of Berlin, denouncing the Navy League as "a union of Conservatives, big industrialists, and financiers." In other respects the organization proved a huge success. A membership of a million, five thousand local branches, with all cabinet ministers and reigning princes of Germany as patrons; a head office with a staff of forty persons (some of them drawing a thousand marks per month), give some idea of the League's success. Its activities were expressed in thousands of meetings, three-quarters of a million copies of a paper *Die Flotte* (The Fleet), and dozens of pamphlets in editions running into millions—all of them designed to accelerate the rate of Germany's naval expansion. Nothing was said about the sales of armour-plate and naval ordnance, but care was taken to assign a particular aspect of the naval policy of expansion to every section of the German community; the middle classes were told about the future growth of overseas trade, the educated classes about coming German world supremacy, and the working class about increased employment and wages in the shipyards.

Under the first Navy Act of 1898 a building programme of twenty-seven battleships was authorized. Their construction was to be spread over six years, during which time no further increase of naval strength was to be asked for. But the very next year the press organs of the industrialists, Krupp's *Berliner Neueste Nachrichten* and Stumm's *Post*, opened a violent campaign for the shortening of the six years' building programme and for a second Navy Act. The pretext for this campaign was provided by an Anglo-German dispute over Samoa, but the real reason was betrayed by the opposition Agrarian *Kreuzzeitung* in an article on trade; the shipyards lacked work! The pseudo-patriotic agitation was, of course, eagerly taken up by the Navy League, which sent out a confidential circular, dated November 30th, 1899, calling for a series of special meetings to stimulate public understanding for the necessity of a stronger fleet.

The purpose of this "understanding" was clearly shown in a widely circulated leaflet: "We need elbow-room at sea. As no one will willingly grant it, we must ourselves take what is our good right, *i.e.*, we must build a fleet strong enough to hold envious foes in check." Germany was rich enough, so the closing sentences read, to let her fleet cost quite a lot of money if necessary.

Unfortunately the results were not those hoped for in Essen. The text of the circular found its way into the press and called forth a storm of indignation over this barefaced attempt to disguise business interests as politics. Corruption in the Navy League was violently attacked and the pressure proved too great for Schweinburg, who, disowned and discarded by his Essen paymasters, was compelled to resign from the Navy League and from the *Neueste Nachrichten.* Nevertheless the firm intensified its underground campaign in an attempt to prevent the Navy League from being turned from a publicity bureau into an ideological association. The task was no easy one, but it was finally done and a rival organization promoted by Berlin professors soon collapsed from lack of funds.

Even the departure of Schweinburg did not cause any slackening of the agitation for increased armaments, as a powerful supporter had now rallied to the cause. Young Emperor William II eagerly adopted the industrialists' slogan for a second Navy Act, and in his famous speech in Hamburg ("Germany's future lies on the water") he urged Tirpitz on when the Admiral was hanging back for tactical reasons. In the spring of 1900 the Reichstag adopted the second Navy Act and the six years' building programme was modified by doubling the quota of battleships and placing a burden of five billion marks on the German people.

But the appetites of the armament firms on the Ruhr and on the Saar were not yet sated. In 1899 they had pressed for more naval orders to enable them to tide over a period of relative slackness at the turn of the century. As soon as the effects of this slackness wore off they reverted to their original tactics and in March, 1901, barely nine months after the passing of the second Navy Act, they set up a clamour for an accelerated programme of naval construction, *i.e.*, for a third Navy Act. Once again the leading part taken by Krupp in this agitation was easily proven.

In a letter dated December 3rd, 1901, the new president of the Navy League, Prince zu Salm-Horstmar, asked Tirpitz to comply with a request made by various party leaders to induce the Government to hasten the rate of naval ship-building, that it might thereby aid depressed German industry and do away with the prevailing unemployment.

Salm-Horstmar later declined to remember who these "various party leaders" were who caused him to write the letter, but he admitted that "they were gentlemen connected with the industrial interests of Western Germany and included Krupp." But this time the efforts of the insatiable contractors proved fruitless. Simultaneously with the writing of this letter a series of five leading articles in Krupp's *Neueste Nachrichten* suggested a modification of the programme of building, according to which a number of ships to be laid down at regular intervals up to 1917 should be put in hand forthwith. Tirpitz, who feared the opposition of the Reichstag, had in the meantime made the embarrassing discovery that the contractors who were again clamouring for Government work had been indulging in shameless profiteering from earlier contracts. His attitude stiffened and he declined to entertain the suggestion.

The disgusted armament manufacturers immediately dropped their agitation and the extraordinary appropriations for the Navy League fell from 412,000 marks in 1900 and 170,000 marks in 1901—to 410 marks in 1902. The publicity organization, now superfluous, was put on starvation rations.

The bitter disappointment of the armament firms over the refusal of the third Navy Act can be appreciated in the light of their fantastic profits—running into many millions—from the first two Navy Acts.

From reliable figures supplied to the Budgetary Commission of the Reichstag, the cost of armour plate for the battleships authorized amounted to 274 million marks, exclusive of the cost of the guns and the cost of the light cruisers. As the profits made by the manufacturers of armour plate in these early years of naval expansion have been definitely proved to be at the rate of over 100 per cent., their total amount exceeded 170 millions, shared by Krupp and the Saarland armour king, Stumm. In the case of Essen this works out at an average rate of at least five millions per annum, guaranteed for sixteen years—for armour-plate alone!

There was now no lack of orders, and the goodwill of the Emperor was doubtless responsible for the fact that the Germania Yard was the only private shipyard which participated in the building of every future class of German battleship. Within a period of fifteen years the Krupp yard built nine battleships, five light cruisers, thirty-three destroyers, and ten submarines—an astound-

FRITZ KRUPP
1854—1902

Wide World Photo

DESK OF FRIEDRICH ALFRED KRUPP, WHO DIED IN 1902. ARTICLES ON THE DESK WERE HIS AND REMAIN UNTOUCHED.

ing record in view of the dissatisfaction so continuously and vociferously expressed by the Navy League.

Germany's fatal naval policy brought golden showers down over the Essen firm. The German naval estimates did not present a complete picture of their extent, as Krupp did not confine the favour of his latest products to his own country. In accordance with the usual practice of armament manufacturers, he proceeded to sell his armour-plate all over the world. He demonstrated its strength on the range at Indianhead before Carnegie, at Shoeburyness before the British naval authorities, and at the Admiralty Works before the Imperial Russian Navy. The demonstrations proved so successful that the leading armour-plate works in England, the United States, France, Russia, Austria, and Italy acquired licences under Krupp patents covering the production of armour. The inclusion of France among these countries is particularly interesting in view of bold attempts made by Krupp biographers to prove that he did not do business with the "hereditary enemy." In this case Paris even got preference, as a French licence was sold and French engineers were undergoing special instruction at the Essen works before any other country, Germany included, took out a licence.

The exact terms on which the various foreign licences were ceded were never disclosed, but from American sources ("Merchants of Death" by Engelbrecht & Hanighen) it would appear that Bethlehem Steel and Carnegie secured licences for Krupp armour-plate production on "terms providing for a substantial licence fee plus a royalty of about 45 dollars per ton." The terms for European licencees would hardly be lower. Krupp thus obtained large cash payments and an interest in the output of foreign armour-plate manufacturers, constituting an important item of his revenue. The results attained by the propaganda campaign conducted by the Navy League and by the Krupp press organs were twofold; they led to increased German armaments and these, in their turn, provoked larger foreign ones. The fruits of this latter aspect of Essen policy were reaped in the Great War when the fleets of Britain, France, Italy, Japan, and the United States, protected by Krupp armour-plate, fought against Germany.

The precise extent of the profits reaped by the Essen firm during this period of intensive arming will never be ascertained until the private archives of the firm of Krupp are open to inspection, but they can be pretty closely estimated by the staggering growth of the Krupp fortune during the corresponding period. According to "The Prussian Millionaires' Annual," published by State Councillor Martin, the taxable fortune of Fritz Krupp was as follows:—

In 1895	- - - 119	million marks
" 1897	- - - 129	" "
" 1899	- - - 148	" "
" 1902	- - - 187	" "

This indicates an increase of fifty per cent., or nearly seventy millions, within seven years, but this record is surpassed by the growth of the personal income of Fritz Krupp, which trebled itself in that time, viz.

In 1895	- - - 7	million marks
" 1897	- - - 9	" "
" 1899	- - - 13	" "
" 1902	- - - 21	" "

The influx of millions was not solely due to the volume of orders but to the fabulous margin of profits extorted by contractors during that time. Tirpitz himself, in the course of his evidence during the inquiry into State control of the armament industry, admitted that from 1899 to 1902 overpayments for warship building were made until such time as an improved system of accountancy at the Admiralty allowed of their being checked. The extent to which the State was exploited by Krupp and by Stumm was proved by the effect of the later costs for shipbuilding on Krupp's income.

Between the years 1901 and 1905 the annual cost of new naval construction (which remained constant as regards the number of ships but increased in respect of tonnage), fell from ninety-three to eighty-four millions, and at the same time the Krupp income suddenly dropped from twenty-one to ten millions per annum. This astonishing coincidence was later made the subject of Parliamentary inquiry concerning profiteering by the firm of Krupp and finally exploded all legends about the "moderate price policy" of the Essen works.

KRUPP, MEMBER OF THE REICHSTAG

The man in whose name such business expansion was achieved was, in contrast to his old-fashioned father, a modern *grand seigneur*, a real prince of industry. In whichever of his three seats—Hügel on the Ruhr, Sayneck in the Rhine Valley, or Meineck in Baden-Baden—he held his court, it was always crowded with the social *élite* of the Empire. The most shining fact of his colourful days was his friendship with the young Emperor and although he maintained his father's traditions by declining to accept a patent of nobility, numerous other honours were heaped on him. William II created him a Privy Councillor, an honour conferring the title of "Excellency" and nominated him to both Upper Chambers of Parliament—the Prussian State Council and the House of Lords. The Emperor referred to the wealthy industrialist as his friend and came to stay with him nearly every year. "Lehmann has come to collect his rake-off" was the disrespectful comment of the workmen when they saw the Imperial Standard flying in the Ruhr Valley.

Apart from their mutual interest in the armaments complex the two young men had much in common. Each was conscious of his own latent weakness and strove to master the problems confronting him. In both cases this led to a hopeless scientific and artistic dilettantism, which only senile historians describe as genius.

In a semi-despotic State like the German Empire the court and retinue of the monarch were of particular interest to the great industrialists and armament makers. The fat contracts which officially emanated from the ministries were in reality awarded by the Court. Parliament was of secondary importance, a mere theatre for the display of opposition by pacifists and occasionally by competitors. It was useful to be represented there, but not by any means essential. When the history of Schneider-Creusot states: "They never failed to secure election to the Chamber of Deputies," it is characteristic of the wholly different status of a great business house in a Parliamentary democracy. The Krupps exercised their influence through the Court and amongst the higher bureaucracy, including that of the army and navy. Their occasional attempts to enter Parliament were inspired more by local political rivalries than by direct business interests.

The septennial elections of 1887 fought on the issue of the Army Programme, brought Fritz Krupp his first defeat as a candidate for Parliament. In 1893 the same issue arose. For although nearly twelve thousand millions had been spent on national defence since the end of the Franco-German war and the strength of the army had been trebled, the Reichstag was asked to approve a further increase of sixty thousand men, equivalent to an additional annual expenditure of a hundred million. When the increase was rejected by the Left majority, William II dissolved his insubordinate Parliament.

In Essen the pro-military Right Party attempted to induce Krupp to renew his candidature. It was no easy task. Fritz Krupp was painfully aware of his own limitations and had no desire to expose them to public criticism, but higher aims—the securing of a majority for the military programme and the fight against the uncompromising Centre Party—induced him to overcome his reluctance. He made but one stipulation: As a prominent employer of labour he did not wish his name used in notices and in the press, so the party conference decided to hold no election meeting or press campaigns on behalf of the candidate.

A "silent" election campaign was inaugurated in which the Krupp partisans shamelessly exercised illicit pressure of all kinds. In the Works men opposing them were threatened with dismissal. The foremen advised their men to take their ballot paper from the "right" distributor at the polling station if they wished to avoid unpleasant consequences. The upshot of this reign of terror was that Krupp secured 19,484 and the Catholic candidate 19,447, but as the Socialists secured five thousand votes a second ballot became necessary. This time Krupp polled 25,000 votes against 22,000, winning from the Catholics with Socialist help.

Krupp's friends were jubilant and staged a great demonstration, which included a torchlight procession and free beer and wine in the local taverns with subsequent brawls and drunkenness. Unfortunately for Krupp the validity of his election was challenged by the Centre Party, who prepared a detailed memorandum on the irregular methods by which it was secured. An inquiry was held, but as usual in the case of members of the party in power, its results were sterile and the elections of both Krupp and of the armour king Stumm (who marched his constituents to the polling booths in columns) were declared valid.

As chief of an armament concern wishing to keep in the good graces of all parties, Krupp avoided adherence to any particular political group. He joined the Empire Party—a small group with reactionary tendencies—as an independent member, never made speeches, and seldom attended committee meetings. Only rarely was he conspicuous in cases where his particular interests were concerned. In 1893 he voted in favour of increasing the peace establishment of the Army—which meant bigger contracts for his firm. On the division for the repeal of the Act against the Jesuits he abstained from voting. As an industrialist favouring the "open-door policy" to stimulate export trade, he opposed the increase of the duty on imported rye. He joined a minority deputation which congratulated Bismarck on the occasion of his eightieth birthday. The majority of members declined to take part because of Bismarck's anti-Socialist laws.

He also voted for Representative Liebknecht's impeachment for *lèse majesté!* In all questions having any bearing on his industrial interests, the Representative Krupp was extremely alert.

The following figures illustrating the phenomenal expansion of the Essen firm's business during this period are worthy of note:—

	1896.	1899.	1902.	
Rothschild - -	216	266	139	millions
Krupp - - - - -	122	148	187	"

While the greatest banking house in the world was curtailing its superhuman activities, the assets of the Essen firm increased by fifty-six millions within six years. What was the comparative amount spent by Krupp—in emulation of Schneider, Skoda, Vickers, and other armament firms—on his vaunted "welfare" work? Did even a small part of these millions wrung from the taxpayer find its way into the pockets of his workers?

After an interval of fifteen years Fritz Krupp embarked on another extended scheme for the provision of workmen's dwellings. A total of 1900 houses in the city of Essen and 1200 at the various other works were erected. They showed considerable improvement in style as compared to the earlier dwellings, including, as they did, modern blocks of flats and garden suburbs. As they were let at the usual neighborhood rentals it is difficult to see why these housing schemes should be presented as part of the renowned welfare policy of the firm, in leaflets and on its stands at exhibitions. The capital expenditure on these housing schemes must, in the light of the vast programme of expansion undertaken by the firm, be regarded as a necessity, and it has been established from official sources that the net return on this investment, after making allowance for all overhead expenditure and taxation, was 2.5 per cent. The firm did not incur any risk whatever on their property, as the dwellings were never empty and tenants could not fail to pay the rent as it was deducted from the payroll. Add to this a factor on which the firm preserved a discreet silence, . . . the rise in land values. The fact that Essen had grown from a little country town into the metropolis of the Ruhr Valley should give some idea of the profit made by Krupp on the land values of his housing estates.

Another point which should be noted is that the tenant was allowed to remain in residence only while employed at the works. If he left or was made to leave, which might be the fate of any man holding "wrong" political opinions, he had to vacate his house the same day. Many a man was compelled to endure being sweated and exploited through fear of being turned out into the street at the first murmur. As 8000 out of a total of 25,000 men employed at the Krupp works in 1901 were housed on the firm's estates, the rentals derived from the latter were undoubtedly an asset in the "welfare" of the firm! Tenants were rigidly supervised in regard to the political nature of the newspapers they read and even as to whether they burned political pamphlets unread. The supervision was exercised by a body of inspectors who were entitled to enter the men's houses at any time in order to see that the numerous regulations laid down for their conduct and mode of life were strictly observed. These visitations led to a condition of mental slavery, making the Works dwellings worse than prison fetters.

The "Altenhof" deserves special notice. In their descriptive pamphlets and books the firm were always proclaiming how well their old employees were cared for. Judging by the wonderful descriptions and illustrations, the world might imagine Essen as a kind of social paradise, where every aged or disabled Krupp employee could spend his declining years free of care. However, this pretty picture must not be examined too closely. At the outset this particular housing estate comprised 100 dwellings, in 1903 their number was 186, and in 1912 the firm announced that the "Altenhof" Estate for Aged and Disabled Employees had a total of 450 dwellings. By this time there were 3000 disabled persons and widows to provide for, so that only one out of every seven could obtain the benefit of any special care. As the total number of employees increased to 70,000 during the period in question the number of aged persons to be cared for increased accordingly, while only a pitiful total of 450 dwellings was provided for their accommodation. Of course they were only allotted to the most deserving persons, at the sole discretion of the firm.

Other prominent features of Essen publicity were the various "funds." These were Insurance Funds into which contributions were paid both by the employees and by the firm; the Pensions Fund, the Widows and Orphans Fund, the Sick Fund, the Families Fund, and the Loan Fund. As a rule these funds were administered on the lines of the Pensions Fund, *i.e.*, in such manner as to deprive contributors of any rights opposed to those of the firm. Membership was compulsory, but contributors forfeited all claims and contributions in the event of their leaving the Krupp employ. Protests by aggrieved victims of this system soon attracted attention from the press, in Parliament, and in the law courts. Subsequent investigations revealed that this vaunted welfare organization did not comply even with the minimum requirements prescribed by law. Under compulsion exercised by an unfavourable legal verdict, the firm had to alter their regulations to remove at least the grossest of the injustices they imposed.

The annual expenditure incurred by Krupp on welfare work of all kinds, in so far as this was laid down by law, were—

Sick Fund - - - - -	51,349	marks
Workmen's Pensions Fund	905,963	"
Staff Pensions Fund - -	660,844	"
Families Fund - - -	14,815	"
Other Funds - - - -	181,256	"
Total - - - - -	1,814,227	marks

These figures are quoted by Krupp admirers and are based on official data. The total annual expenditure on welfare work by the Essen firm may therefore be put at about two million marks. Contrasting this with the figures extracted from "The Millionaires' Annual for 1902," the following comparison is obtained:—

Krupp fortune - -	187	million	marks
Annual income - -	21	"	"
Annual increase of fortune - - - -	20	"	"
Annual expenditure on welfare - - -	2	"	"

These figures speak for themselves. Krupp gave his army of employees only five per cent. of the sum represented by his income plus the capital growth of his fortune. In view of the colossal profits derived from monopolies, the naval construction programme, and the heavy profiteering indulged in by the firm, this contribution is a very modest one, more especially in the light of the advantage of a permanent body of resident and highly specialized technical workers. The cost of assuming the guise of a "welfare firm" in place of the unsavoury one of war contractors living on the blood of others does not appear to have imposed a heavy burden on the House of Krupp.

But the world in general believes in the semblance of things, and overlooks the fact that Krupp was a warmonger and an industrial reactionary. The head of the Essen Administrative Board was Councillor of Finance Jenecke, a fierce opponent of the Utopian ideal of a standard working day, who hoped to make the Works a bulwark against all social-revolutionary ideas. The actual part played by him and his chief in wrecking the Socialist policy initiated during the early years of William II's reign has never been fully ascertained. But Jenecke himself has stated that German industry owed its uninterrupted supremacy to Fritz Krupp, as its position would have been rendered hopeless if the greatest, most powerful, and wealthiest industrialist in Germany had ever given the slightest indication of departing from the principle that the factory proprietor is, and must remain, master in his own house.

CONCERNING THE RECOIL CYLINDER

A history of weapons of war written for non-military readers might be interesting, as it would give us some idea of the amount of metallurgical, constructional, and manufacturing work which has to be carried out on a piece of steel to fit it for that function of our civilization for which it is designed—the destruction of humanity. Such a history would also throw light on the pitiless struggle of great and small interests which make the arms industry, as we know it, such an utter farce.

Towards the end of the century ordnance underwent a technical revolution, as far-reaching as the invention of solid-drawn breech-loaders. Since the first adoption of quick-firing magazine rifles, the defects of the "fixed" field guns in general contemporary use were becoming increasingly apparent. At every shot the recoil caused the gun to "buck" and to throw back the gun-carriage, rendering it necessary to "point" the gun after each round. This slowed down the firing continuity badly, although the use of smokeless powder made it possible to remain on target continuously. The best guns of those years possessed a slower rate of fire than that of the primitive mortars of Gustavus Adolphus, a truly appalling performance for an engine which, since that time, had become so much more expensive to produce.

Theoretically it does not seem difficult to find means of absorbing the recoil, as neither consideration of heavy weight nor of rigid attachment to the ground arises in the case of field guns. But—the idea was already in the air—would it not be possible to introduce a device between the gun and its carriage that would take up the force of the recoil and apply this stored-up energy to the purpose of putting the gun back to its former position? This principle was the basis on which the earliest non-rigid mounted guns were designed; but the new idea was not developed sufficiently in their case, as the travel available for braking was too short, so that the force of the recoil was merely reduced and not absorbed, a doubtful advantage in view of the increased weight involved by the braking devices.

Experts were unable to progress beyond this point, which is precisely the point at which a humble and obscure engineer named Konrad Haussner began some novel experiments. After a series of fantastic projects he developed the idea of the "long recoil cylinder." In 1888, while employed by Krupp as a designer of vehicles, he prepared an interesting memorandum which he submitted in November of that year to the Head of the Gun Factory. The memorandum had fourteen pages of print with diagrams and exact calculations, fully developing the theory of a field gun with a sufficiently long recoil cylinder, whereby the gun ran back on its carriage as if in a cradle and was brought forward again after the recoil by means of compressed air, as Haussner first suggested.

Some weeks later the young inventor's memorandum was returned to him with the remark that his design was not considered suitable for use in the field. The decision was given by the brilliant Gross, who thereby proved that he was becoming stale, unreceptive and opposed to disconcerting innovations—just as were the Works, themselves, frozen in their monopoly rights. Possibly the immediate superior of the inventor may have influenced him against "the young man who wants to teach us how to build guns although if he knew how to calculate at all he would know that no such mounting could possibly be built."

The verdict of self-complacent bureaucracy carries the day. The disappointed inventor resigned from his Essen job. He improved on his original design and applied for a patent, which was granted to him on April 29th, 1891. He subsequently admitted that his patent claims were practically worthless, as they did not cover the actual system of the long recoil cylinder which he had invented but merely the particular form of construction to which he had applied it. After enlisting financial support, Haussner built an experimental gun and submitted it to the Ordnance Research Board in Berlin for test. After the latter had administered the customary snubs reserved for unknown inventors, they evinced some interest; whereupon a licence agreement was entered into with Gruson who received from the Board an order for an experimental gun with a long recoil cylinder. This appears to have broken the ice, and the now hopeful inventor joined the employ of the Magdeburg firm who appeared to be really interested and immediately began the construction of two guns.

It was at this stage that the Gruson works were taken over by Krupp. A severe blow to Haussner, now compelled to deal with the people who had previously turned down his ideas. As the guns were completed, however, Essen could not refuse to submit them to practical firing tests. These were duly carried out at Meppen in March, 1894, in the presence of Prussian officers, but under singularly unfavorable conditions.

The curved spade on the trail of the carriage was obviously too small for the hard ground at the firing point of the range, but the management of the works specifically forbade any alterations to be carried out. The results of the firing were accordingly poor and to the relief of the works representatives the Ordnance Research Board's officers reported unfavourably. The president of the commission, Colonel Reichenau, called out "Away with the monstrosity!"

The inventor made a final effort to break through the wall of official prejudice and technical ignorance by preparing another design which provided for a recoil spring, and two months after the fiasco at Meppen he submitted his proposals to the engineer acting for Gross during the latter's absence on leave. A few days later, without even an opportunity of giving any explanation, his papers were returned to him with the curt order, "Take your picture away again." The expression "picture," added Haussner, meant in technical parlance "that what was illustrated was rubbish."

That was the verdict of Essen on the greatest artillery invention of those years, which it now rejected for the third time. The idea was finally discarded by the firm who, as legal successors to all the Gruson business, returned the Haussner patent and dropped the whole proposition.

But somebody else did not drop it—Heinrich Ehrhardt, the chairman of the Rhenish Metal and Engine Works. Just as Louis Baare was the most serious competitor of the Krupp works of Alfred's day, so Ehrhardt was at the close of the century. Starting as a penniless apprentice (and colleague of young Skoda), Ehrhardt had become a brilliant engineer and creator of the Rheinmetall works in Düsseldorf. His competition was particularly distasteful to Krupp, inasmuch as it attacked his special products, *i.e.*, the manufacture of ordnance. Using lightweight seamless tubes, produced by a patented process, for making gun carriages, the Ehrhardt works had attained a degree of technical prominence which their chief, an exceptionally able man of business with modern ideas, utilized with the utmost success.

Ehrhardt eagerly grasped the opportunity of acquiring an invention which appeared likely to shake Essen's monopoly. He did not hesitate to communicate with Haussner, although the latter was still an employee of his competitor. An agreement was concluded under which Ehrhardt agreed to construct an improved gun from new drawings. In October, 1896, the strangely restless inventor became manager of the Ehrhardt Vehicle Works in Gotha. He applied for a further patent on the construction of the improved gun with carriage in which Düsseldorf patent tubes were used. In the spring of 1897 the gun was demonstrated to Prussian ordnance experts who were inclined to be sceptical, notwithstanding the excellent results of the test to which it was put, owing to the said experts' failure to appreciate the importance of the new invention irrespective of the unavoidable defects of this first model.

In 1897 military authorities in Berlin were about to make a final decision as to the type of field gun with which it was proposed to re-equip the German Army. Trusting the opinion of Krupp and the Ordnance Research Board (which was under his influence), it was decided to adopt only the rigid gun, pattern 96 (with folding spade). The greater part of the orders, amounting in all to 140 millions, were placed in Essen.

This was the moment for which France had been waiting. Ever since the defeat of 1870-71, the more alert

Photo from European

MAIN ADMINISTRATION BUILDING OF THE KRUPP WORKS AT ESSEN.

KRUPP WORKS IN ESSEN, 1933.
(Aerial Photo)

military authorities in Paris had kept a secret and careful watch on Haussner's work. This fact would never have become known but for the disclosures made by the newspaper *L'Illustration* on February 6th, 1915, in the excitement engendered by the first months of the war.

"General Mathieu, who was Director of Artillery at the Ministry of War in 1890, was informed through the usual sources that a distinguished German engineer named Haussner had constructed at the Krupp works a gun with a long recoil cylinder. After a series of tests the firm of Krupp were stated to be ready to manufacture the new gun. The General, who was a keen student of human nature, sent for Major Deport, at that time Director of the Puteaux Gun Factory, and asked him whether he could construct a gun on the principle of the long cylinder recoil. Major Deport, who knew something about the matter, replied after due consideration that he was prepared to try and solve the problem and in 1894 he submitted a field gun, capable of firing up to twenty-five rounds per minute, to the Minister of War, General Mercier."

One might have imagined that while Major Deport was thus engaged, German artillery experts were equally busy. But this was not the case, as German artillery had not only failed to make any progress during that time but had even lost ground by adopting a totally erroneous policy. The information furnished to General Mathieu which led to the creation of the 75 mm. gun was, astounding as it may seem, quite incorrect.

The engineer Haussner had certainly designed a gun which was produced in Essen. But the doubtless purposely unfavourable results of the tests to which it was subjected led the firm of Krupp, only too glad of an excuse to reject a proposition diametrically opposed to their own practice, to turn the invention down and dismiss Haussner, who departed for South America to seek his fortune. Moreover, the patent maintenance fees in France remained unpaid, Haussner's French patent was invalidated, and although quite unknown, made open to public inspection. Messrs. Krupp lost the finest opportunity ever offered to them and thanks to their invincible stubbornness lost it for all time.

While France was equipping her army with guns of the latest pattern, Krupp supplied the all-too-trusting military authorities in Berlin with 140 millions' worth of field guns of a pattern which, according to the ordnance expert von Perbandt, "was obsolete even before it came into the hands of the troops."

Essen seemed to have gone completely blind. Even after it had become known that France was adopting the new system, Krupp published his "Firing Test Report 98," in which it was stated that "complete absorption of the recoil involves forms of construction totally unsuitable for use in the field." Three years later Lieut.-Colonel Leydhecker (retired), a senior employee of Krupp's, published a further unfavourable criticism in the *Swiss Artillery and Engineering Journal*. It is hardly possible not to agree with General Wille when he asserts in his book "Ehrhardt Cannon" that Krupp's attitude on the matter of the long-recoil cylinder did not, even in 1901, "differ from the opinion held by him from 1895 to 1898." But at last came a rude awakening to the prolonged slumber of the self-complacent monopolist firm. This time it was Switzerland, contemplating new armaments, which gave the impetus towards the new technical ideas. After long tests by the Artillery Commission the Federal Council, in March, 1901, recommended equipping the Field Artillery with Krupp cannon. That would seem a final blow to the long-recoil cylinder. But the Swiss Legislature refused to accept the proposition put forward by short-visioned military men and did not conceal its doubtful attitude towards the rigid field gun. Thus were contracts in the millions definitely lost to Krupp.

The effects were far-reaching, and artillery experts throughout Europe began to take notice when they heard that Ehrhardt had sold 108 of his despised guns to the United Kingdom and had just had an order for 132 from Norway. For the first time Krupp had to admit defeat at the hands of a German competitor, and after a brief period of reflection he declared in favour of the long opposed recoil-cylinder gun. In order to cloak the technical barrenness of his late policy he announced that the works had been experimenting on the construction of a recoil-mounting with roller system. But this fact, as General Wille remarked, appears to have been remarkably well concealed from the rest of the world.

Essen now remembered the long-forgotten Konrad Haussner. His ideas had been contemptuously rejected ten years earlier, but was he not then an employee of the Works? And did not all inventions, including those rejected, automatically belong to the firm? Was there not such a clause in Haussner's service agreement? Basing their claims on these facts, the Essen firm began a lawsuit against Ehrhardt contesting the latter's right to the invention. An application for the cancellation of Haussner's patent of December, 1896, was fought out before the Imperial Patent Office and the High Court, and in both cases the proceedings showed up the Essen firm in unfavourable light. Krupp shifted his ground continually, first stating that Haussner's drawings were made in the firm's office, and then asserting that his statement only applied to the general principle of the invention. Moreover Krupp's statement that he deferred action in respect to Haussner's second patent until he had established the fact that the patentee's conduct was irregular, was incompatible with the omission to take earlier action in connection with Haussner's first patent. Krupp's inconsistency proved that he was trying to build up his case from information given by witnesses under oath, and that he really had had no case of his own. The court was not called upon to inquire into the question as to what particular features of the invention were developed at the Krupp works but to decide whether the invention as a whole had been made there. In default of proof to that effect Krupp's case must fail and the patent be declared valid.

Düsseldorf's victory for the patent rights of the long recoil cylinder was by no means the end of Krupp's bitter fight against Ehrhardt. There was actually little to choose between the business methods of either party, as is proved by Ehrhardt's treatment of the man to whom he owed his greatest triumph. If Haussner imagined he was to reap a rich financial harvest through the commercial exploitation of his invention, he was speedily undeceived and before long was engaged in disputes and lawsuits with Ehrhardt. The courts finally recognized Haussner as the actual inventor of the long recoil cylinder, but the ingenuity of the self-advertising Düsseldorf industrialist enabled him to take all the credit for producing the "Ehrhardt gun." In his memoirs he actually wrote: "Herr Konrad Haussner was one of those people who, notwithstanding all I did for them, displayed nothing but ingratitude towards me." He even went so far as to accuse Haussner of stealing the Ehrhardt ideas without producing any proof whatever in support of the accusation. Haussner's treatment by the armament interests was typical of the business ethics characteristic of these gentry. Krupp first treated him with contempt and then tried to infringe his patent; the French armament industry stole his ideas through factory spies; Ehrhardt used his invention to rise to industrial eminence himself and then repudiated him. The genius of the unknown engineer had enriched the magnates by several hundred millions.

In 1903 the position of the German army authorities was unfavourable from a military point of view and indefensible from a commonsense one, by their blind trust in Krupp, who had led them to adopt a type of gun at a time when it was already fit for nothing but the scrapheap. It became increasingly difficult to conceal this painful state of things, and awkward disclosures made in the technical press resulted in an extremely unpleasant debate in the Reichstag in March, 1903. The Socialist leader August Bebel accused the Minister of War of wasting 140 millions by rearming the German artillery with obsolete guns, on the plea that the French artillery had just been rearmed throughout, although in actual fact the latter acquired their new weapons *after* the German artillery had been rearmed, and the new French guns were vastly superior to the German cannon.

The irate War Minister endeavoured to deny the failure of the military authorities as well as of Krupp. He was careful to pay no attention to Bebel's assertion that the Budget Commission had been duped by false reports in regard to French armaments. But truth was on the march, and a very costly truth it proved. Perbandt writes:

"Four years after delivery of the 96 type guns the German Government was forced to make good Krupp's unprogressiveness at a cost of about one hundred million marks for "improvements" on these guns.

Among the many services rendered to the country by the Essen firm as recorded by its admiring historians, no mention is made of the "improved pattern 96 guns"; although this little matter of the wasted hundred millions was never allowed to affect the excellent relations prevailing between the authorities and the firm. In his memoirs Krupp boasts that in 1903 the military authorities finally decided to accept the "Krupp recoil cylinder gun." But he omits any reference to the private visit paid to Hügel by Minister of War von Gossler, just prior to this decision. Krupp was allowed to make money both ways—on the supply of inferior guns and on subsequent necessary improvements. All in all, a cool two hundred millions.

THE "LEAP IN THE DARK"

Writers outside Germany have consistently declared that William II had a financial interest in the firm of Krupp, although there is considerable divergence of opinion as to the nature and extent of this interest. The Emperor is stated to have been either a shareholder or a financial backer of the firm by people who advance no proofs in support of such statements; and all attempts to throw light on the relations of the Crown to the gun works have proved abortive—including attempts by Republican newspapers in post-war Germany to prove their existence.

According to an announcement made by the Berlin Ministry of Finance in February, 1920, following sequestration of the private property of the ex-Emperor, no stocks or shares in the firm of Krupp were found amongst it. But it was also established that on February 3rd, 1908, William II instructed the Deutsche Bank to purchase Krupp shares to the value of 50,000 marks, which shares were disposed of through the Prussian Maritime Trade Organization in April, 1914. This is evidence that the ex-Emperor was interested in the Krupp business, but the extent of his holding was absurdly small. A far more significant fact is the presence on the otherwise highly exclusive board of Krupp directors of the Berlin court banker Ludwig Delbrück, which might have made possible a far more subtle financial contact. Old Alfred Krupp avoided the great banks on principle, but never hesitated to ask for the financial accommodation he required from Emperor William I's privy purse or from the ministries. This system of concealing the sources of their financial support established a tradition to which his successors adhered.

It would however be a mistake to assume that the relations between the Emperor and the firm of Krupp were inspired by any idea of mere gain on the part of the Monarch; they were more likely due to the Imperial policy of concentration, which had as its aim a system of mutual economic and political support. As Eckart Kehr puts it: "The Agrarians (feudal landowners) gave Industry a fleet, and Industry gave the Agrarians a tariff." This was carried so far as to cause ministerial departments openly to invite Krupp and other great industrialists to subscribe funds for propaganda purposes. In a letter dated August 3rd, 1898, and addressed to ten leading industrialists, the Secretary General of the Federation of German Industries stated that he had been authorized by the Home Office to invite their subscriptions to a fund of 12,000 marks for conducting a campaign for a new law relating to industrial labour regulations. Krupp, through his Berlin representative Jencke, subscribed 5000 marks, although he considered the request of the Home Office to be a "somewhat singular one." The latter utilized the funds contributed by Krupp in its publicity campaign for a new Prisons Act.

Times had changed since Minister von Roon rejected a dozen Krupp tenders and the old Prussian bureaucrats had made no secret of their distrust of the bourgeois upstart's commercial machinations. Both Government departments and big capital concerns were now in such close association that it was difficult to draw a clear dividing line between them. The Minister of Labour, Major-General Budde, was a former director of Krupp's associated firm, the Karlsruhe Arms and Munitions Works, and was also on the board of the Belgian arms factory in Herstal. His brother was one of the senior executives in Krupp's employ as were also the younger brother of the Secretary of State and future Imperial Chancellor Bülow, and the son of the President of the Artillery Research Board, closely concerned with the allotment of contracts. As regards the navy, Captain Sack, who had been equerry to young Prince William at the time of the latter's visit to Essen, was by then Vice-Admiral in charge of the Equipment Department of the Admiralty, and dealt with contracts for naval ordnance and ships' armour. Later on he joined the boards of directors of Krupp, of the Cologne-Rottweiler Powder Trust, and the Karlsruhe Arms and Munitions Works. This record had doubtless been achieved through the influence of the Emperor, who also contrived to secure the appointment of his personal A.D.C. von Grumme to the board of the Hamburg-Amerika Line, property of his friend Ballin.

Apart from all this unconcealed jobbery, the friendship of Fritz Krupp with Minister von Rheinbaben, Secretary of State Hollman, and dozens of generals, admirals, and high officials, is worthy of note. In all matters relating to technical supervision, tenders, and questions of finance the representative of the Essen firm invariably had to deal with "a friend of the house" in the person of the Government representative concerned. The historian Hallgarten, after years of research in the German State records, declares that "whole Government departments eventually became Krupp agencies."

The Emperor's interest in the purely business concerns of the Essen firm was amazing. In the Brandt case, which occupied the courts just before the Great War and revealed a few of Krupp's business secrets, the defence definitely established that in 1895 or 1897 a memorandum was circulated in the Ordnance Department in which the "All Highest" gave instructions that Essen should supply a certain proportion of artillery deliveries, after which the balance could be divided between other firms. Contemporary opinion of the friendship between the monarch and Krupp is reflected by the following lament in the Berlin newspaper *Vorwärts:* "The Emperor has gone to stay with Krupp. Ostensible reason: target practice. Probable result: new artillery proposals."

William's eagerness to further the business interests of the firm was by no means limited by the frontiers of his own country. He wrote to the Tsar, that "Nicky" should get his new ships built in Germany, "as our private firms would be delighted to receive orders." In his memoirs von Heyking writes: "The Emperor considers the most important duty (of the Minister in Pekin), is to persuade China to order as many ships, &c., as possible from us . . ." Essen could certainly point to the successful achievements of its Imperial agent with satisfaction; they included the construction of the cruiser "Askold" for Russia and of three battleships for China, in the Germania Yards.

Great though these Imperial favours to the friends or partners of the Essen concern might be, the influence exercised by the Krupp business interest on matters of national politics was even more important and far-reaching, especially in regard to foreign policy. The reactionary tendency at home gave rise to a policy of aggression abroad, dictated by commercial and industrial interests (particularly by the interests of the armament firms) which impelled Wilhelmstrasse, during the ten years following the Bismarck era, to pursue its alarming activities in the Far East, on the Bosphorus, and in South Africa. The non-renewal of the Berlin-Moscow Defence Pact, . . . the first step towards the encirclement of Germany, . . . called attention to Krupp's Turkish armament contracts, which for the first time affected Russia's designs on the Straits. The fortification of Germany's eastern frontier with its adverse influence on relations with Russia was likewise not unconnected with Krupp interests. The doubtless well-informed friend of the Emperor, Field-Marshal Waldersee, wrote on the matter in his diary: "The idea of seeking to protect our exposed eastern frontiers by means of fortifications is a sadly unhappy thought. . . . Unfortunately dubious motives affect this policy: the vast profits to be derived from the supply of armoured turrets, gunmountings, armour plate, &c., by the great industrialists lead them to exploit the Emperor's inclinations for their own purposes."

Ample proofs of the evil influence exerted by the Essen firm on the foreign policy of the Empire are contained in various official publications on German diplomacy, *e.g.*, "The Guiding Policy of European Cabinets, 1871-1914."

The consequences of Germany's action in joining France and Russia to oppose the demands of Japan at the Peace of Shimonoseki, which the dismissed Bismarck called "a leap in the dark," were particularly serious. Berlin's partisanship for China was due to the excellent reason that Viceroy Li Hung Chang happened to be a valued old customer for guns, rails, and warships supplied by Krupp, who was the chief German exporter to China. Krupp agents in Pekin were far more influential and important people than the official diplomatic representatives. One of them, Dr. Baur, was appointed chief instructor for the Chinese railways—an ideal post for the representative of a concern manufacturing rails. His colleague Mandl encouraged the hesitating Li Hung Chang to resist Japan in Korea. When at the commencement of hostilities Britain proposed mediation by the Great Powers, Wilhelmstrasse curtly declined the suggestion. Reason for this attitude is disclosed in an official document—a confidential report, of March 19th, 1895, from the Imperial Chancellor von Hohenlohe to William II, which runs as follows: —

"German trade has hitherto remained unaffected by the war. On the contrary our manufacturers, exporters, and shippers have been afforded a good opportunity to do business in deliveries and transport of war materials."

The Emperor approved the rejection of the mediation proposals in consideration of the business the war was bringing Krupp, and duly wrote "Approved" on the margin. But the crushing defeat of Li Hung Chang by the Japanese and the stiff terms exacted by the latter in subsequent peace negotiations at Shimonoseki led to a change in Germany's attitude. Acting in concert with France and Russia her Minister in Tokio, Baron von Gutschmid, intervened and, in marked contrast to his colleagues, made an aggressive speech threatening war. This blow delivered on behalf of Krupp's friend Li Hung Chang proved effective. Infuriated and humiliated, the Japanese abandoned Liautung Peninsula and Port Arthur.

But friendly relations with Pekin were not always consistently maintained. The Essen practice of selling to both opponents led to some unpleasantness during the Sino-Japanese war, concerning which the historian Hallgarten writes: "Krupp has a representative in China, named Mandl. This gentleman is extremely enterprising and with the connivance of two highly placed Chinese officials, one of them a nephew of Li Hung Chang, he sold obsolete guns to the Chinese Government, while at the same time, and with the aid of the same persons, he supplied the victoriously advancing Japanese enemy! This proved too much of a good thing even for the

Chinese Government, who proceeded against their two officials for high treason. The Krupp representative Mandl, worried about his contracts, rushed off to the German Minister, with the request that he bring pressure to bear on the Chinese Government to abandon their intention. But as this was likely to prove an awkward matter, the German Minister declined to intervene. Disappointed and indignant, Mandl immediately reported all the circumstances to Essen. Whereupon the head of the firm, Privy Councillor F. A. Krupp, then frequently in Berlin, mentioned the matter to the Permanent Under-Secretary of State for Foreign Affairs, Privy Councillor Mühlberg, in the course of a morning ride in the Tiergarten. When Krupp himself spoke, Wilhelmstrasse had to obey; the disobliging Minister received a severe official snub, whereupon he sarcastically inquired whether certain reports forwarded by him concerning defects in Krupp guns used in this last Far Eastern war had aroused wrath in Essen. . . . As may be imagined, this retort did not mollify the annoyed Foreign Office, the diplomat's position became insecure and, as he soon afterwards offended German financial interests, he was recalled to Berlin."

The sequel to the Sino-Japanese war brought Krupp further large contracts, including the refortification of Port Arthur. He also became head of the "German Shantung Syndicate" which secured a virtual monopoly for exploiting the mineral wealth of this rich Chinese province. The value of the contracts for rails alone ran into hundreds of millions.

Five years later Germany was repaid for her assistance in a very different coin. During the Boxer Rebellion, the gunboat "Iltis" engaged the Chinese Taku Forts in a combat, which has since been praised as one of the most heroic episodes of Germany's naval and military history. But the various legends concerning it omit to state what the severely wounded Commander Lans, subsequently awarded the *Pour le Mérite* decorations, has to say about it: "We were hit 17 times, most of the shells bursting in the ship and killing and wounding many of my brave men. And the irony of it! All the enemy's guns and shells come from our own country; they are all modern Krupp quickfirers."

Published official German records do not always tell the whole truth. In the 200 pages of Foreign Office records of the period covering the first Peace Conference at The Hague in 1899, scarcely any mention is made of the armament interests, although just from that quarter came the stiffest opposition to the proposal to limit armaments. Only the Emperor in his marginal notes sometimes calls things by their right names. In the course of a conversation with Count Eulenburg (report dated October 23rd, 1898) the Russian Foreign Minister Muravieff referred to the idea of limiting the introduction of technical improvements in armaments and went on to make the following malicious comment:

"One aspect of the matter which might, perhaps, be of special interest to H.M. the German Emperor is that a limitation means a reduction in the huge numbers of men employed by Krupp and others on work connected with new experiments."

The Emperor's comment, taken down by his Secretary of State Bülow, was "How is Krupp to pay his men?" These words might sum up Germany's attitude toward The Hague Conference. It was not due to any considerations affecting the country's interests that an all-round international reduction of armaments was opposed, but to the imperative need of finding new outlets for the activities of the Essen works, expanding at a phenomenal rate. The same considerations applied to export trade in arms, which would have been paralyzed by disarmament. Krupp's best customer, Sultan Abdul Hamid, addressed an anxious inquiry to the German Ambassador von Marschall as to the latter's opinion concerning the possibility of Turkey disarming. Marschall reported—his reply marked "Confidential"—on November 4th, 1898, to Imperial Chancellor Hohenlohe:—

"I did not feel justified in raising any false hopes, which were not reflected in the existing state of things. When, in the course of my audience, references to supplies of ammunition and rifles were made, I ventured the opinion that, pending a realization of the new world order visualized by the Russians, the old adage, '*si vis pacem, para bellum*,' was an unanswerable dictum for humanity."

Germany was, of course, not the only power to wreck the more serious proposals discussed at The Hague. Both British and French attitudes were similarly inspired by armament interests. None of the official documents appertaining to the Peace Conference contain anything comparable in clarity with the historic comment, entered in manuscript, on the records dated June 23rd, 1899: "To hell with all the resolutions! William R.I."

So long as the aims of Germany's various financial groups were unified, the Government could pursue a definite course. But it became hard to do so in 1900 during the Boer war, where the interests at stake were for the first time not parallel with the frontiers but cut across the various States. What was termed the "betrayal of the Boers" was actually the vacillation of Wilhelmstrasse under pressure by rival cliques. One group, headed by the Berlin Chamber of Commerce, the Bochum Union and the Wörmann Line, were interested in the building of a railway from Pretoria to the coast, *i.e.*, they were pro-Boer. Another, represented by the Deutsche Bank, sided with Cecil Rhodes, whose gold mines and diamond fields gave promise of rich dividends. This group included the firm of Krupp and in Essen the popular romantic predilection for the "kindred Boer people" found little response. Moreover, orders coming from the Transvaal in 1897 and 1898 had not left any doubt as to the fact that the best customer was in London.

As the Boer Republics were prevented from importing war material by the blockade at sea, strict neutrality would mean no war material supplied to their opponents. But shortly after the beginning of the war there were rumours from Essen of a mysterious contract for 45,000 shrapnel shells, the drawings of which did not bear the customary signature. At first the Works management attempted to deny that the contract came from the United Kingdom—which it obviously did—but the public outcry was so great that the Government was forced to move in the matter. The officially inspired *Norddeutsche Allgemeine Zeitung* of January 13th, 1900, referred to it as follows: "Statements have repeatedly appeared in the press that the firm of Krupp in Essen are engaged on the completion of an urgent contract for a large quantity of steel projectiles for England, and the question has been raised as to whether the declaration of strict neutrality made by the Government of Germany in regard to the war in South Africa admits of the supply of German-made war material to either of the combatants. We are informed that the competent authorities have replied to the question in the negative and Messrs. Krupp were accordingly requested, immediately after the appearance of the above press statements, to refrain from the dispatch of any arms, guns, munitions, and other warlike stores to either combatant."

The pressure of public opinion thus vehemently aroused on behalf of the Boers, Krupp promised not to supply war material. It involved no actual loss to him as other more circuitous ways of supplying his British customers were already arranged. When later reports stated that London was negotiating with Krupp for 240 quick-firing guns, an immediate denial was issued; the report referred to older pattern Krupp guns from the arsenal of a South European (?) power. At the same time Berlin newspapers reported that Italy (?) had placed orders to the value of fifty million marks with Krupp—a remarkable action on the part of a country whose financial state at that time could hardly permit such expenditure. It was all merely a way to circumvent the neutrality restrictions and Fritz Krupp had not broken his word in the strict letter of the law; he maintained German neutrality and still earned huge profits. Shortly afterwards it became known that the firm of Ehrhardt had likewise supplied several batteries of guns to the British Government without any attempt at concealment, whereupon the Essen works truculently announced that they would in future disregard the Government's request and would supply guns and arms to England at all times.

Contemporary public opinion failed to appreciate the full significance of the effect of these Krupp transactions in the Far East and in South Africa, but fifteen years later Japanese troops shouted "Vengeance for Shimonoseki!" as they advanced to attack Kiao-chow. President Wilson justified the American supplies of war-material to the Entente by citing the precedent set by German firms during the British blockade in the Boer war.

CAPRI

The personally timid and bashful Fritz Krupp, who found it hard to say "no" to anybody, had, under the influence of his bureaucrats, turned into an uncompromising reactionary. Their energy triumphed over his yielding weakness. He avoided personal contact whenever he could and spent most of his time away from Essen, in Nice, San Remo, or Capri, where the multimillionaire found an atmosphere widely different from that prevailing in the sober and workaday North. He was always surrounded by hosts of visitors who vied with each other in flattering the man on whom they sponged. The extent of their servility is illustrated by the following statement of an eye-witness named Meisbach: "Some years ago Dr. Schweninger recommended Mr. Krupp to lie on his stomach for an hour after meals, which recommendation the latter duly carried out. All his male guests followed his example in order that their host should not be bored by having to do it alone."

Krupp played the part of an old-style *grand seigneur* in dispensing hospitality to such toadying guests, but there was a ceaseless under-current of gossip and intrigue beneath this façade of gold. It had its beginnings while Krupp was still showing preference for the more restricted, purely German social circles in which he moved at the turn of the century, but it became more definitely critical when the unrestrained atmosphere and more free and easy habits of life in the South surrounded him.

At this particular time the head of the Berlin C.I.D., von Tresckow, was especially active in discreet investigations of a confidential nature entrusted to him by Court circles, and his memoirs contain some interesting details. He tells of the special section in the Berlin Criminal Files devoted to cases of sex-perversion blackmail, which flourished in prudish pre-War Berlin. In following these shameless blackmailers the criminal Police came upon facts which induced their Head, Baron Meerscheidt-Hullessem, to keep a special carefully guarded secret file. For among persons known, or suspected, as sex-perverts were many of the most brilliant names of Court circles. Hohenau, Bodo von Knesebeck the Empress's private secretary, the Court Chamberlain von Oppen, Prince Frederick Henry of Prussia, the Military Commandant of Berlin, Count Kuno von Moltke, and the Counsellor of the French Embassy Lecomte. All the names, titles, occupations, ages, and addresses of the persons concerned were noted in the card index together with remarks relating to the causes which had brought them under suspicion, the people with whom they consorted, and the sources from which they were blackmailed.

The uneasiness caused amongst the elect by the possible revelation of this unusual record of the Emperor's friends can readily be imagined. Meerscheidt-Hüllessem anticipated a scandal, and gave directions in his will that this sinister card index should be delivered to the Emperor after his death. Von Tresckow's account of the discharge of this duty states:—

"The person entrusted with it, an old friend of the deceased, dispatched the sealed package containing the card index together with a covering letter to the Emperor, to whom it was in due course submitted by the then President of the Civil Cabinet Lucanus. The Emperor did not open it, however, but merely returned it with the curt remark, 'It's probably some police matter, so send it to the Commissioner.' "

The attempt to enlighten William II failed and the secret documents found their way back to the closely guarded confidential safe, but they were none the less carefully perused by von Tresckow. While reading them, he made the incredible discovery that this record of social criminals included the name of . . . Fritz Krupp!

This unfortunate episode in the private life of Krupp might have been overlooked, in so far as its tragic sequel had no bearing on his activity as an arms industrialist, but for the fact that the reckless accusations of slander and intellectual murder to be shortly hurled at his political opponents called for a minute investigation of all the circumstances connected with the "Krupp case."

The manuscript notes on the card index were, as von Tresckow remarks, very inadequate and he supplemented them by results of his own investigations. These included information supplied by Conrad Uhl, proprietor of the Bristol Hotel, where Krupp always stopped when in Berlin, concerning the singular interest displayed by Fritz Krupp in young waiters whom he even used to send to the hotel from Italy. Uhl had numerous letters from Krupp relating to the treatment and supervision of his *protégés* and as Uhl finally objected to such interference with the internal management of his hotel, he protested to Krupp when the latter again came to the Bristol. Krupp was angry and moved to another hotel. But next time he came to Berlin, he went back to the Bristol. Uhl noticed that he never had his wife with him in the hotel even when she came to Berlin. Husband and wife stayed at different hotels. Uhl drew his own conclusions in regard to the personal habits of Krupp.

Soon afterwards von Tresckow interrogated a suspicious character, arrested in a police raid on one of the resorts of habitual criminals, because of the huge diamond ring he was wearing. The man stated under cross-examination that the diamond was a present from Krupp in whose service he had spent many years as a footman. Krupp's personal relations with this man accounted for the gift.

Krupp's biographers lay great stress on his generosity to people in his personal service and on the lavish gratuities which he habitually bestowed on waiters, cooks, porters, and pages in the hotels in which he put up. But, as one writer puts it, "one cannot help regretting that this generosity did not extend to the employees in his works!"

Fritz Krupp spent the spring of 1902 in Capri, which he had frequently visited, and where he was well known. He had purchased a "holy" grotto—the Hermitage of Fra Felice—on the island and had improved its accessibility by having a road cut to it through the rocks. He installed a caretaker, clad in the garb of a Franciscan monk, in the grotto. In one of his gayer hours he founded a men's association, the "Order" giving the members Italian names and private keys to the grotto.

Capri priests began to murmur "sacrilege." The tennis courts which Krupp had laid out were said to spoil the beauty of the island. The visitor from the North was fond of the sea and, although he chartered a small steamer for his own use in the previous years, he came in his own yacht on this occasion. With Dr. Salvatore of the Marine Zoological Research Station of Naples, he spent much time on the Bay of Salerno, fishing for new specimens of deep sea fauna. Krupp did not possess a villa of his own at Capri, but occupied a princely suite in the Hotel Quisisana, the proprietor of which, Cavaliere Serena, was the Syndic of the Island. They were close friends and the industrialist often entertained the hotel proprietor in his castle of Hügel.

Fantastic rumours of the foreign nabob's generosity attracted crowds of artists from all over Italy. Krupp made numerous and tasteless purchases—he is stated to have bought no fewer than twelve busts of Dante; his hospitality knew no bounds. The swarms of visitors at his table included members of his elect "Order" distinguished by their tie-pins, representing miniature gold-mounted projectiles, presented to them by Krupp. Vast quantities of Capri wine were a feature of all-night orgies in the luxurious apartments. Sometimes Krupp provided displays of fireworks for his guests and he also arranged for concerts and recitals by violinists, trained and retained by himself, in the barber's shop opposite Pagano's Hotel.

Curious tales concerning the habits of the generous host continued to circulate and he could not have helped hearing about them. But he disregarded them to such an extent that the tales and rumours ultimately led to serious charges of moral perversion against Fritz Krupp. The latter utterly failed to appreciate their gravity and was even completely indifferent when informed that photographs of himself as a partaker in the orgies were being circulated.

Matters reached such a pitch that the Italian authorities began to take notice and to make discreet inquiries in Capri. News of their action naturally reached Berlin, where von Tresckow received first-hand information from a C.I.D. colleague in Capri to the effect that Krupp's conduct there had passed the wide limits usually extended to matters of this kind in Italy. The Roman newspaper *Avanti* had already printed an article on it and the German press was bound to refer to it shortly. Tresckow suggested to his superior officer, Commissioner of Police von Windheim, that he should inform the Emperor, a personal friend of Krupp.

The Commissioner demurred at this suggestion as he feared that the Emperor would take exception to such a course, when there was, as yet, no evidence to substantiate the charges against his friend.

At the end of May Krupp suddenly left Capri. The Rome authorities strenuously denied that he had been requested to leave, but it is more than likely that he received a polite hint to do so. The Italian press dealt with the "Krupp case" vigorously and fully, and a definite accusation supported by a mass of regrettably detailed evidence was levelled against the German industrialist by the Neapolitan newspaper *Propaganda.* Other newspapers, including the Roman *Avanti,* took up the matter and added to it further information which they had gathered. Public indignation was aroused to a high pitch, for although liberal pre-War Italy regarded sex perversion as entirely a private matter, the corruption of minors was considered a grave offense subject to severe punishment.

After the matter had been pursued for some weeks in the Italian press, without evoking any denial on the part of Krupp, von Tresckow's fears were realized; the German press took the matter up. On November 8th the Catholic *Augsburger Postzeitung* printed a warning article from Rome: "Unfortunately the case closely concerns a great industrialist of the highest reputation who is intimately connected with the Imperial Court." Krupp and his directors still remained silent—as if they knew nothing of the matter. If they imagined that a topic which was engaging the attention of the world's press could be passed over silently in Germany, they soon received a rude awakening. On November 15th, 1902, the Berlin daily paper *Vorwärts* contained an article headed "Krupp in Capri." The accusations made in the Italian press were fully reproduced and all the relevant names given, although details were omitted. The tone of the article was moderate and concluded with a humane suggestion to repeal the sections of the penal code which prescribed penalties for those thus abnormally afflicted.

But the publication of Krupp's name made the affair a true sensation. Immediately after the appearance of the article a series of urgent telegrams was exchanged between Berlin and Essen. Did Fritz Krupp realize that he could not procrastinate any longer? Did the administrative board wish to clear the whole matter up? A few hours after the appearance of the newspaper a suit for libel was filed against it by wire. The authorities decided to take up the matter officially, and ordered confiscation of the newspaper containing the offending article. The order was carried out in a manner unusual even in Imperial Germany. The police raided the office of the newspaper and even searched the desks of members of the Reichstag. In the provinces, where numerous local papers had reprinted the article, confiscations were also carried out and in some cases copies of offending newspapers actually removed from subscribers' residences.

The machinery of the law having been set in motion, it was to be anticipated that a thorough investigation and clearing up of the case would follow. Practically the whole of non-Socialist public opinion was convinced of the libellous character of the attack and a complete exoneration of the greatly maligned industrialist was confidently looked for at an early date.

Dark clouds hung over the ivy-clad walls of Castle Hügel in the Ruhr Valley in this third week of November of the year 1902. The legal adviser of the firm, Korn, furnished the first authorized explanation of the affair, now several months old. His statement read like a fairy tale and was to the effect that Krupp, through his friend Cavaliere Serena, had become involved in a Capri election campaign, the Syndic declaring that the future attitude of the island's wealthy patron depended on his victory. His political opponents had recourse to "scandalous accusations," in which they endeavoured to "cast base suspicion on the incidental and absolutely harmless meetings of Mr. Krupp with his Capri and German friends, in the most despicable manner." Furthermore, certain persons in Capri felt aggrieved because the foreign benefactor "had rescued a number of humble folk out of usurious clutches." The Socialist press retaliated with the words: "All these activities occupied the gentleman to the extent that he could not find time to answer a wire from his Essen workers, describing pay cuts and other abuses. If Krupp's employees had been Capri folk he would probably have freed even them from the usurers' clutches."

Then the Milan newspaper *Secolo* proceeded to prove that the rumours concerning Krupp were known long before the election in Capri. The debate as to the veracity of the Italian charges continued and a notice in the Krupp works announced that "criminal proceedings against the authors of the libel and slanders have been instituted."

The person chiefly concerned remained silent. After spending three days in Kiel he returned to Hügel on November 19th. Did he realize his position and was his prolonged silence an acquiescence in the coming *débâcle?* Were there none amongst the many who had enjoyed his hospitality to offer him advice and assistance? What did the telegrams from the Emperor's intimates say? What was the attitude of the administrative board, from which Hans Jencke had significantly resigned some weeks previously? Slowly but surely the suspicions crystallized into definite charges against this publicly honoured man who had just returned to his home. On the evening of the 21st he dined alone with both his daughters. He pulled himself together in front of his children and did not betray any signs of uneasiness. Subsequent reports that "he was particularly cheerful" are obviously unworthy of credence. On the contrary he was not cheerful that evening, but inwardly desperate, when he closed the door behind him—never to open it again.

On the evening of the following day, the official German news agency reported:

"Villa Hügel November 22. His Excellency Privy Councillor Krupp died at three o'clock this afternoon.

Death was due to a stroke which he suffered this morning at 6 o'clock."

All Germany was stirred by the death of the greatest industrialist of the Empire, the multi-millionaire and head of the Essen armament firm. Barely forty-eight years of age, he died just as he was facing infamous charges, the libellous character of which the courts had not yet had time to establish. What had really happened? The question which rose to the lips of all was put by the *Kölnische Zeitung*, "Did knowledge of his own guilt impel him to take his life?"

There was good reason to believe that official versions of the cause of Krupp's death were untrue. The nearest approach to the truth was, perhaps, disclosed in the local Essen press, derived from "absolutely authentic sources" and which stated: "On Friday night His Excellency Mr. Krupp felt unwell and instructed his valet, who was alone with him, to send for his doctor in the morning. When Dr. Pahl Samstag arrived at Hügel next morning the servant, in reply to his question, said that Krupp was sleeping well but was snoring very loudly. The doctor went to the bedside and saw at a glance that the snoring was really a death rattle." This is a somewhat different story from the official version of two strokes, at 6 a.m. and towards noon, clearly intended to leave no room for conjecture. The singular attitude of those in charge of the Hügel household was quite in keeping with it. The coffin was sealed a few hours later, affording no opportunity for an examination of the body. At the private memorial service for relatives, friends of the family, and members of the administrative board, nobody was permitted to view the body of the deceased. In view of the growth of generally accepted opinion that

Krupp committed suicide, this procedure appears to have been absolutely inexplicable. Was there any sign of a bullet wound? Had any poison (narcotic) caused outward tell-tale changes to appear?

The obvious thing would have been to publish a medical autopsy report to dispose of the rumours once and for all. But this was not done. Only vague references to a confidential autopsy report, signed by competent medical authorities, were made in the press. The *Berliner Lokal-Anzeiger* announced that the report would not be published as it constituted important evidence in the prosecution of the *Vorwärts* in view of the connection between the libellous attacks and the stroke which followed them.

Krupp's widow, Margarethe, had been undergoing treatment for three weeks in a mental home in Jena, under the supervision of Professor Binswanger. She returned to Hügel the day of his death.

Her stay in the mental home was not unconnected with the events of the last few months. The abnormal peculiarities of her husband had actually driven this able and energetic but puritanically reserved woman to take a serious step. The *Munchener Post*, so well informed on all Krupp matters, wrote on the subject:

"In desperation Mrs. Krupp appealed to the Emperor, who was looked upon as a valued friend of the family. The lady's representations affected the Emperor so deeply that he advised the adoption of drastic steps, including a possible application declaring Krupp incompetent, in order to preserve the Works, the property, and honour of the family. This became known to Krupp's personal friends, whereupon one of his intimates, Admiral Hollmann, moved heaven and earth to avert the fate threatening Krupp. His wife was described to the Emperor as an hysterical woman, frequently incapable of controlling herself through jealousy, and as being in urgent need of prolonged treatment in some establishment for nervous disorders. Mrs. Krupp was accordingly placed under medical observation and it was only due to the vigilance of a certain medical man that she was not certified insane and immured in a mental asylum. Nevertheless she remained under observation and supervision, and on the very day of Krupp's death a meeting of mental experts to pronounce on her state was to have taken place at Hügel."

The biographer Meisbach confirmed the above statement which was further elaborated in the newspaper *Zukunft*, normally strongly pro-Krupp. The paper declared: "The Emperor, wishing to avoid a scandal, insisted on Krupp leaving Capri immediately and on his never returning there. . . . Krupp thereupon began to fear that inquiries into his sanity might be made." What the issue of the whole affair might have been if Krupp's sudden death had not put an abrupt end to it, is difficult to say. But now that the man was dead, fallen in a conflict to which the participation of the Government's Socialist enemies had given especial weight, the case outgrew the limits of a mere family tragedy. It became a political issue, a matter of State, and William II was quick enough to see it and to drop the attitude of detached reserve which he had hitherto maintained.

During the night of November 26th the body was carried to the "Old House" within the Works, the third Krupp to be laid there in state. Next morning the funeral services were held, and Essen had never seen the like before. Behind the coffin, in the uniform of a General and wearing the Black Eagle decoration, walked the man who was the cynosure of all eyes, of the thousands of other important mourners . . . Emperor William II. Thoughtfully he walked through the black-draped street between the rows of flaring torches. He knew the full significance of the monarch's appearance behind a citizen's coffin, but he had come there "to hold the German Emperor's shield over the House and the memory of the dead man," as he himself announced solemnly. The streets were crowded and police could not keep back the throngs. Finally a squadron of Düsseldorf Hussars, riding ahead of the procession, cleared the way.

The graveside address was delivered by Bishop Klingemann. He referred to the last words said to have been spoken by the deceased before he lapsed into coma: "I go in peace with all men, free from any feelings of hatred and bitterness against those who have wronged me so deeply." Was it by chance that this sentence had been so freely quoted in so many different forms? Klein-Hehemann, the editor of the local paper, reproduced it as "I depart this life without hatred and bitterness and forgive all those who have injured me." The memorial plate bore the words: "I forgive all my enemies." What was the truth? Was it not expressed in the first announcement of Krupp's death, issued by the administrative board, that he died "without regaining consciousness"?

The Emperor remained silent at the graveside. But on returning to the main railway station he delivered an extemporized address to a carefully selected audience in which he referred to the close ties of friendship binding him to the deceased and to his family, to the high qualities possessed and ever displayed by the deceased, and to the infamous calumnies which had caused him to fall a victim to his unassailable integrity. The Imperial orator concluded his peroration with wild abuse of those who were seeking to besmirch the fair name of the dead man.

Official accounts of the address suppressed the most extravagant expressions used in it.

William II was determined to make the death of his friend an excuse for drastic action against the Socialist opponents of the State. On December 5th, while in Breslau, he received a deputation of workmen who expressed their "deepest and respectful gratitude" for his intervention in the Krupp case. Fears of repressive measures had induced several thousand workmen to append their signatures to the declaration. The same procedure was followed in Essen, Bochum, Kiel, and in the Magdeburg Gruson works, although in the last-named establishment two men refused to associate themselves with the movement and declined to sign. Their service of sixteen and twenty-two years respectively did not prevent their consequent summary dismissal. This led to energetic protests in all the cities from which the "Imperial addresses" emanated—revealing the forcible measures of moral compulsion used for this purpose. The protest brought results. In the Reichstag Imperial Chancellor Bülow declared a complete disavowal of the method used to force loyalty: "If the workpeople wish to make such declarations, they must be made freely and not under pressure, if they are to be of any value. Manifestations inspired by outside pressure are, I consider, absolutely worthless."

Even more painful was the failure of the press campaign instigated by the Emperor to put over his legal interpretation of moral murder.

At first even the more critical organs of the press associated themselves with the rehabilitation of the honour of the deceased and a spirited endeavour was made to prove that Krupp was the victim of mistaken identity. A report from Rome stated that the *Giornale d'Italia* was in a position to prove that local inquiries in Capri had resulted in the issue of a warrant for the arrest of a German citizen accused of grave offences against minors, no fewer than ten different complaints being made of his conduct. Krupp was not on the island at the time the alleged offences were committed.

The paper went on to state that the real culprit was a painter named Allers, an individual of doubtful antecedents, who owned a villa in Capri. Krupp's friends let loose a storm of indignation against him with a view to diverting public attention from the dead industrialist. They omitted mention of the fact that Allers was a friend of the Bismarck family. Allers was alleged to have obtained huge sums from his patron Krupp by way of blackmail. The case against Allers was, however, abandoned for lack of evidence and his legal advisers announced that the warrant was withdrawn and the charges dropped. Few of the German newspapers took the trouble to publish this vindication of Allers' honour.

In the midst of this sensational press campaign, which had by now acquired the character of a political agitation against the anti-armament Socialists of the Left, a veritable bombshell was dropped on December 15th, some weeks after the Emperor's speech, by an official communication to the *Vorwärts* which read as follows:

"The issue No. 268 of the *Vorwärts*, published on November 15th 1902, contained an article headed 'Krupp in

Capri' which made grave allegations against Privy Councillor Friedrich Krupp.

"A writ for libel was applied for by wire on the same day by Mr. Krupp and written application to that end was lodged at the local Public Prosecutor's office on the following day.

"Under instruction issued by the Royal Court of Justice I on November 17th, the issue No. 268 of the *Vorwarts* was ordered to be confiscated, which instruction was duly carried out.

"On November 22nd Mr. Krupp died.

"The plaintiff was therefore unable to put in a sworn statement in answer to the attack made on him.

"He repeatedly declared it to be his firm intention to prove the baselessness of the allegations made against him, as is shown by the proceedings for libel which he instituted.

"Under these circumstances the widow of the deceased, Mrs. Krupp, has stated that, in view of the certainty of her husband's innocence, she is desirous of ending the publicity involved by the proceedings relating to the deceased. For this reason she does not wish to pursue the criminal prosecution of the originators and publishers of the libel any further.

"I have therefore decided that it would not be in the public interest to proceed with the charge of criminal libel, which is accordingly withdrawn.

(signed) "Dr. Isenbiel,
"Public Prosecutor."

The effect of this decision by the legal authorities of pre-War Germany was amazing. A leading newspaper had just proved the necessity on legal grounds alone of pursuing the charge of criminal libel, even of homicide, against *Vorwarts* and now this decision to drop it was announced! German newspapers of those days presented a picture of complete bewilderment. In so far as they ventured to make any comments at all, they guardedly expressed mistrust. The management of the firm of Krupp made an effort to mitigate the effect of the official decision by amplifying it:

"Mrs. Krupp has intimated to the Public Prosecutor that she is not prepared to express a wish for the continuance of legal proceedings, and is quite ready to leave the matter to his discretion."

This is obviously in direct contradiction to the announcement made by Dr. Isenbiel, who stated that "Mrs. Krupp is extremely desirous of ending the publicity involved by the proceedings," whereas the management said that "she is quite ready to leave the matter to the discretion of the Public Prosecutor." The general confusion produced by this contradiction was so great that countless interpretations of their implications were put forward.

It was generally assumed that the abandonment of the prosecution had been instigated by the Emperor himself, who had at last appreciated how ill-suited was the Krupp case for publicity purposes. Popular opinion was voiced by Baron von Fechenbach-Laudenbach: "After due consideration of all the circumstances we must conclude that the Public Prosecutor did not abandon investigations on his own initiative but under instructions from a higher authority." This view is confirmed by statements published in the well-informed Munich daily paper already referred to: "The Prussian Minister of Justice has had a lengthy conference with Public Prosecutor Isenbiel on the Krupp case and the circumstances connected with Krupp's sudden death. The decision made by the Public Prosecutor was the result of this conference." Personal intervention of a Minister in an affair of this nature would be unthinkable in semi-despotic pre-War Germany, where the Ministers of the Crown were the personal representatives of the monarch, unless it were made with the approval of the Emperor.

The Krupp affair continued to engage public attention for a long time and after it had been dealt with by the press and the courts for several months it finally reached the Reichstag, where, on January 20th, 1903, Representative von Vollmar was called to order by President von Ballestrem for referring to it, because of the fact that the death of a private individual in tragic circumstances must be considered irrelevant to Parliamentary debate which concerned other matters. When von Vollmar attempted to show that the case in point was made a pretext for the Emperor to deliver speeches attacking the political party which he represented, the President ruled that he was out of order, and von Vollmar was finally compelled to drop the matter. The fact that an event on which the interest of the whole world had been focused had to be passed over in silence in the German Parliament, was a revelation of the latter's weakness in face of the powerful influence exercised by the armament industry. The utmost indignation was expressed on all sides at the high-handed procedure adopted by the President in the matter and his conduct was criticized even by the Conservative press organs. On January 23rd Ballestrem resigned his office as President of the Reichstag and a few days later the Emperor nominated him to hereditary membership of the Prussian Upper Chamber. William II was at pains to prove that his solidarity with the House of Krupp was not to be shaken by any public accusations.

Popular misgivings regarding the Krupp case remained unallayed for some time and persistent rumours that the accused multi-millionaire was not really dead continued to be circulated, elaborated by the usual stories about a mock funeral—to which some colour was lent by the fact that a strict watch was kept on the private burial ground of the Krupp family for a long period. Reports in the press to the effect that travellers in America—or even in Palestine—had seen the "Cannon King" alive continued to appear at intervals for some years.

PART V

THE KRUPP A.-G.—THE LIMITED COMPANY

The Giant Grows—Parliament and Armour Profiteering—Conflict with Le Creusot—Prince Consort von Bohlen—The Great Panic—"Kornwalzer"—The Downfall of Ehrhardt—The Duel of the Giants—On the Eve—The War—The Conversion—Secret Arms—The Crisis You Need—Forward Over Graves

THE KRUPP A.-G.—THE LIMITED COMPANY

THE GIANT GROWS

VAST shops in Essen, Annen and Magdeburg resound to the thunder of smelting furnaces, rolling mills, power hammers, armour-plate, gun foundries and forges. Iron mines on the Rhine and in Northern Spain, together with a dozen collieries, deliver their output directly to the furnace mouths by the firm's own ships and railways. Huge steel hulls rise from the slip-ways of Kiel shipyard. On the three artillery ranges at Meppen, Dülmen, and Tangerhütte, the newest and heaviest guns fire day after day—40,800 rounds per annum, more than in a whole sector of the front line in the Great War. Every morning an army of 24,000 men streams into the workshops and offices.

The time is 1903 and this industrial undertaking the largest on the Continent: a property worth 273 million marks, with its latest annual profits returned at twenty-one millions net. All this had been left by the last male representative of the house. The nominal heiress was his sixteen-year-old daughter, Bertha Antoinette, but the real ruler of the Works and of the Hügel residence was the widow, Margarethe Krupp.

She was transplanted overnight from the nursing home where her mental state was being studied, to the leadership of an industrial kingdom. The courteous epicure of yesterday, content to allow things to take their own course, had become a disillusioned and energetic woman. She had had ample opportunity to observe the state of things, to appreciate the strained contrast between the princely household at Hügel and the men in the works; here luxury and nepotism, there bureaucratic torpidity threatening the continued existence of the carefully built-up monopoly. The ruthless exposure of the firm's affairs to the glaring light of publicity—carefully guarded privacy had been a tradition handed down by old Krupp—must have been particularly painful to a woman of her reserved nature. She had seen all this and much more with great concern. The sudden departure of Jencke, after many years' service as head of the firm, and that of the creator of new ordnance, Gross, was a significant sign of the times. Margarethe Krupp took the reins with feminine caution which her natural stubbornness occasionally turned to stark harshness. Her rule of the Works was marked by the scanty and sterile nature of the information available concerning internal affairs of the business. As in the days of Helene Amalie, the future of the firm in its critical hour was in the hands of a woman married into the house of Krupp, of whose spirit she now appeared as living embodiment.

The will of the deceased (the contents of which were never disclosed although they were the subject of dispute) directed that the firm was to be reorganized as a private limited liability company in order to regularize relations between owner and administration. No surrender of proprietary rights was contemplated and no Krupp shares were ever to be dealt in on the stock exchange. The house of Krupp considered themselves the sovereigns of an industrial kingdom with strictly regulated rights of succession long before the idea of an industrial inheritance in the modern sense took shape. 159,996 shares of 1000 marks each were allotted to young Bertha, and, in order to comply with legal requirements, the remaining four shares were distributed to close friends whose discretion not to pry too closely into confidential matters, such as the firm's relations to the Crown, might be relied on. The conspicuous difference in the valuation of assets and the amount of issued capital (113 millions) constituted the original "first reserve." The legal obligation to publish annual balance sheets must not be allowed to endanger any business secrets of the administration of armament industry. Krupp accountancy remained unintelligible, in spite of repeated complaint from the commercial press concerning such concealment.

Appointments to the board of directors and management were made with due regard to existing contracts between the State and the firm. Old-Prussia's statesmen had displayed reluctance to nearer relations with the firm, Minister of Commerce Maybach and Postmaster-General Stephan having declined the Cannon King's pressing invitation to join him. But, now, Minister of Railways von Thielen and Rear-Admiral Barandon had no scruples in joining a firm to whom their excellent connections with the Government and the Admiralty might be useful. The board of directors was also strengthened by the Emperor's court banker Ludwig Delbrück, and by Otto Budde, brother of the Minister of Labour, Major-General Budde. At a later date they were, as already stated, joined by Friedrich von Bülow and by Admiral von Sack. Apart from members of the family and other close friends, the board of the Krupp A.-G. in 1903 included a retired cabinet minister, a retired admiral, a retired captain, a court banker, seven highly-placed civil servants, nine engineers, business men and legal experts, all drawing annual fees of over 100,000 marks, and it was "patronized" by the paternal friend of the house, William II himself, even though he cautiously remained away from Castle Hügel—doubtless in consequence of the family incident.

Barely eight weeks after Fritz Krupp's death his widow placed a contract for the construction of a great steelworks in Rheinhausen, the little town on the left bank of the Rhine nearly opposite the mouth of the Ruhr. Plans for this scheme had been prepared a long time before, when the exposed situation of Lorraine in case of war made it advisable to arrange for works elsewhere. Land had been acquired, a blast furnace and jetty built. But the unfavourable site of the original works, now surrounded by a sea of houses of the Ruhr cities, undermined by countless coal galleries and with no direct access to a waterway, rendered the transfer of important sections of the works from Essen to Rheinhausen a matter of necessity. It had taken the short-sighted bureaucrats in charge fifteen years to realize this, and only when the matter had become urgent was it taken in hand by order of Mrs. Margarethe Krupp. Three years later the Rheinhausen Friedrich-Alfred Works with up-to-date blast furnaces, coking plant, rolling mills, and the first large Thomas steelworks, were put into service. The adoption of the Thomas process (incidentally years after its adoption by other German plants) enabled the Krupp A.-G. to move up from the ranks of manufacturers specializing in the production of certain classes of steel into those of the great steel producers. In contrast to the obsolete and jumbled conglomeration of workshops in Essen, Rheinhausen was a really modern Works, capable of turning raw material into the finished product "in a single heating" thereby eliminating wasteful cooling down and reheating.

The Krupp works had, however, developed to a stage which afforded small scope for individual leadership. Mrs. Krupp displayed great energy in suppressing waste and internal friction, but she was neither willing nor able to curtail the prevailing cumbersome system of administration. The foreman of many years' service and the experienced engineer or manager who looked upon their employment at Essen as their life's work, were the prevalent type of employee in the Krupp concern; a type capable of rendering much conscientious service but lacking in foresight and intuition. This explains the reason why, in addition to the recoil cylinder already mentioned, all the remaining inventions relating to ordnance made during these years (*e.g.*, the single-lever breech mechanism, independent gunsights, explosive shrapnel, the mechanical time fuse, &c.) were not only produced outside Essen, but were actively opposed by Krupp until such time as the latter was compelled to purchase them at great cost.

This technical torpidity and internal red tape found its counterpart in the external syndication of Germany's industry, then forming a number of large trade associations to stifle irksome competition. In 1904 a number of small syndicates amalgamated to form the German Steel Works Association, Krupp with an output of nearly one million tons taking second place. State Councillor Dr. Völker, appointed by the Government to allay public anxiety by conducting an inquiry into the dangers of this gigantic syndicate, was secretly approached by its competent executives with an offer of a post drawing an annual salary of 100,000 marks. He accepted promptly without any compunction regarding his official status.

This combination of staunch business management and bootlegging principles in its application was particularly displayed in the main activity of the firm, *i.e.*, the ordnance department. The Essen Employees' Club was ruled by the sour spirit of the Protestant Sunday School, but the managers' offices of Krupp agencies were reminiscent of Chicago. The Krupp representative was hand-in-glove with the foreign statesman and the common grafter, imperial military attachés and disreputable spies of all kinds, with both benefactors and hangers-on. The tentacles of the Essen armament octopus stretched around the globe and no single metropolis or potential sphere of interest was overlooked. A "constitution" of the Krupp industrial kingdom was laid down.

The firm's outside interests were represented and safeguarded by representatives (agents) of the following four classes:—

1. Principal Agents, termed "Plenipotentiary Representatives of the Firm of Krupp."

2. Agents, who in view of the more restricted and relatively less remunerative scope of the activities allotted them were, subject to certain conditions, permitted to undertake representations for firms other than Krupp, whereby, of course, the business interests of the last-named firm were to be in no wise prejudiced.

3. Sub-Agents; these were appointed for individual territories, cities, or small business circles within the sphere assigned to the Principal Agents and Agents referred to in Classes I and II.

4. Sub-Agents who bought the firm's products on their own account.

This system was applied to the whole world. Krupp agents were active from Tokio to Madrid and from St. Petersburg to Buenos Aires, urging each country to buy guns and armour for use against its neighbour, to whom other Krupp agents were simultaneously offering similar guns and armour. The roughest work was of course done by the lesser men, particularly by the interesting group of Sub-Agents "buying on their own account." They were recruited from among foreign business men and private persons possessing useful connections with the Governments concerned and made use of their friends and relations to sell cannon—generally of obsolete pattern—to them. The machinations of the gentry involved in such transactions were not always recognized as easily as in the case of Li Hung, nephew of the Chinese Viceroy, or that of Colonel Nicaise, who as Belgian Minister of War endeavoured to find employment for his brother-in-law Brialmont, in Essen. There was also some unpleasant notoriety in the case of the President of the Italian Chamber of Commerce in Paris, Mario Cresta, who, during the Lybian war addressed patriotic meetings of his countrymen while selling fifty Krupp mountain guns to their Turkish enemy.

The prestige of the Essen firm's name was reflected in the high-sounding titles borne by the aristocrats amongst their army of representatives, "Plenipotentiary Representatives of the Firm of Krupp," which were purposely intended to resemble those of Ambassadors and Ministers Plenipotentiary. These senior representatives of Krupp accredited to foreign Governments were important people possessing valuable contacts. In Vienna it was Herr von Ficzek, a friend of the Rothschilds; in New York J. B. Satterlee, a relative of J. P. Morgan; and in Copenhagen Herr Madsen, the future Danish Minister of War. The fact that the firm represented by these gentlemen had at all times the ear of the Imperial Chancellor and of the Emperor rendered it essential for Germany's diplomatic representatives to pay every attention to their inquiries and wishes. Any kicking against the pricks—as when the German Minister in China, von Brand, had the temerity to point out technical defects in Krupp products—led to an abrupt termination of their diplomatic careers. Fortunately, failure to defer to the wishes of the Essen firm in matters of business was a very rare occurrence. The most efficient and successful of Germany's diplomatic representatives abroad were devoted friends of Krupp. For instance, in Constantinople, on which the attention of Germany's pre-war foreign policy was so assiduously directed, the palace of the Plenipotentiary Representative of the Firm of Krupp was adjacent to the German Embassy, and the Ambassador was at pains to be a good and attentive neighbour. The Krupp spy service functioned perfectly and passed many a titbit to the Embassy. On June 13th, 1903, the Chargé d'Affaires, von Wangenheim, reported to Imperial Chancellor von Bülow, in a document classified as "Secret":—

"According to what I hear locally and from Bulgaria, the Prince's Government are making a great show of humbleness and pacifism, due to their lack of preparedness for war. Meanwhile, the Government are in secret treaty—as I am informed by the local Krupp Agency—for the purchase of arms and munitions in Vienna and Paris."

One was certain to be well informed through Krupp sources and deserving of the Emperor's marginal comment, "Excellently written! Congratulate Wangenheim on my behalf."

Such tips as these, of course, had to be repaid in kind. The German Embassies were gradually turned into auxiliary agencies of the Essen firm—not merely to deal with foreign competition. Schichau shipyard and Ehrhardt discovered to their cost how closely the German official machine, from the Emperor down to the last junior Secretary of Embassy, was identified with Krupp. Complaints about this state of affairs, even when taken up in the Reichstag, proved unavailing. Albert Helms stated: "I know that one of our leading firms takes particular care, when transacting business abroad, to keep the German Government in the dark about the proposition concerned, as the diplomatic representatives might have their attention drawn to it, and this attention might merely benefit Krupp."

At ordinary times the Krupp emissaries contented themselves with simple representation and business espionage, but if there was any prospect of a contract running into millions the agencies became active. Subtle contacts and crude bribery were brought into play. The most modest member of Parliamentary or military commissions, even the actual staffs of ordnance centres, became greatly courted personages. Comparative gunnery tests became real Wild West shows, with burnings of opponents' store-houses and unexpected refusals by the railway authorities to carry dangerous goods; during the tests officials proved themselves to be fanatical adherents to some particular system, and if despite all this the opposing competitor made a good showing, the results were unexpectedly disallowed by the commission. It was not easy to throw light on this murky competition, and in the few cases where it could be done, as in South America and on the Bosphorus, it was solely due to business conflicts developing into major political issues. As a general rule the international armament interests concerned preferred to maintain silence and although it was war to the knife, care was taken not to betray any information which might prove injurious to both sides. Statements that "means other than business methods" were resorted to in order to influence the decision of a government in South-East Europe, were countered by some mysterious reference to an article "in No. 6008 of the leading press organ of the country concerned." Or Ehrhardt issued a denial of the rumours "emanating from a certain quarter" of gun explosions in Austria, but gave the obviously false assurance that no implication concerning Krupp was intended thereby. These were the considerately delivered foil thrusts made under strict rules of the game, to which neither of the opponents took exception.

The arms industry boomed throughout all Europe in 1902 owing to the success of the recoil cylinder principle. Big customers of the year included Switzerland, Holland, Denmark, and Rumania. Shortly after the conclusion of the Boxer Rebellion, which had cost Germany so many lives, the Chinese Minister in Berlin went to Essen, where the firm of Krupp accepted a large contract for the rearmament of China, regardless of popular indignation recently aroused by the shelling of the gunboat "Iltis" by Chinese guns of Krupp make. The dividends declared by the new limited company were in keeping with their dazzling turnover:

1904–6% 1905–7½% 1906–10%

Krupp's competitors and rivals in the battle between the rigid gun and the recoil cylinder, Ehrhardt's Rheinische Metallwarenfabrik, were less fortunate. Although they were the real creators of the recoil cylinder principle and of the improved breech mechanism, their output was crippled by an unending series of patent actions brought by Essen. In his report for 1904 Ehrhardt complained: "We would to-day be swamped with ordnance work if it were not for certain attacks made on the designs which we have produced at the cost of such heavy sacrifices." While Krupp announced the payment of fat dividends his competitor had to write off a substantial loss. The titan of Essen was slowly beginning to crush his last surviving German rival.

PARLIAMENT AND ARMOUR PROFITEERING

In contrast to Vickers, Schneider, and other great armament concerns of the democratic West, Essen had never attached importance to Parliamentary connections. The board of directors did not include any House member, such as Paasche at Ehrhardt's and Erzberger in the Thyssen concern. Krupp contented themselves with the maintenance of the traditional contacts with Court circles, government departments, and, above all, with the heads of the army and navy.

This looked like involving a day of reckoning when

German public opinion, alarmed by the Navy Acts, began to look askance at the misuse of the blessings conferred on multi-millionaires. It looked for a while as though the ensuing debate on the cost of armour plate and on the ordnance monopoly might arouse the futile German Parliament into a semblance of activity.

In the spring of 1900 the proceedings of the Budget Commission, which were of a semi-private and unrecorded nature, had already given rise to some preliminary skirmishing. The Opposition now attempted to reconnoitre the position in the light of the Government's demands for nineteen battleships and eight large cruisers. Working from data furnished by an expert, who temporarily concealed his identity, the Catholic newspaper, *Kölnische Volkszeitung*, began to reckon out Krupp's profits. In nickel steel alone these worked out at the net amount of 99 million marks in sixteen years, equivalent to a net annual profit of six millions, divided between Krupp and Stumm. Before the Commission even more alarming figures were quoted; out of a total contract expenditure of 279 millions for fleet armament no less than 176 millions constituted net profit. The discomfited Tirpitz refrained from any attempt at denial, merely said he was "unable to give any information." Representative von Kardorff quoted a letter written by the steel magnate Stumm which was revealing, notwithstanding the denials it contained; the cost of the armour plate was "only" 113 millions, of which sum "not more than half" represented the manufacturer's profits. So the profits on armour plate alone would seem to have been some sixty millions, without allowance for those on the armament and ship-building part of the programme—a truly damning admission!

It was also revealed that Essen enjoyed the patronage of the authorities to an extent that virtually precluded competition of other firms; as when the Admiralty required contractors to provide their own artillery range for acceptance tests, which was obviously beyond the resources of a new undertaking. This led to a complete exposure of the Krupp armament monopoly. Public opinion was shocked, the press aroused, and Parliament became alarmed. It was in this mood that the Reichstag met to discuss estimates for 1901 and it determined to put an end to the scandal.

In the Budget Commission on March 1st a member of the Centre Party, Müller-Fulda, opened the attack. Once more figures giving irrefutable evidence were quoted; in the previous year Krupp had arbitrarily raised the price of armour plate, and notwithstanding certain subsequent reductions the State was still compelled to pay one million more per battleship than formerly.

But the real sensation came—from Tirpitz himself! Müller-Fulda's report to the House on the matter read:—

"The Secretary of State caused a document in the English language to be read before the commission by a captain, wherein it was confirmed that we pay four hundred marks more for armour plates than American naval authorities pay their makers for them."

What had happened? Why this attack on the armament contractor by the chief agitator for increased armaments? The Commission was not disposed to confine itself to questions on causes and influences, but concentrated on the exposed issue between those who wished to build as many ships as possible and those who aimed to make a maximum profit. It snatched at the information contained in the mysterious document, which included surprisingly full particulars. The U. S. Navy Board, so it said, had succeeded in obtaining a reduction of 455.5 dollars (1920 marks) in the price of armour plates, to which the contractors had agreed only after Congress had approved a proposal to expend four million dollars on the construction of a national armour-plate rolling plant. The reduced price was 400 marks lower than that demanded by the German contractors, although the American firms had to pay royalties to Krupp and to Harvey.

No clearer proof of Essen's profiteering could be hoped for. Four hundred marks per ton meant an annual sum of three million marks wrung from the State! Nobody dared deny the necessity of putting an immediate stop to this sort of thing, but the question was how? The moral of the United States conflict over prices was too obvious for the Commission to ignore. The Conservative Count von Stolberg, a supporter of the Government Party, moved a resolution to establish a State armour-plate works and the motion was adopted by twenty votes to four.

A storm now threatened to break over the heads of Krupp and Stumm. On March 8th, Müller-Fulda gave the report of the Budget Commission to the Reichstag and this report concluded with the damning statement: "In view of the unprecedented profiteering indulged in by two German armour-plate firms, who have created a monopoly for themselves, it is considered essential either to call for competitive tenders from abroad or to establish a State-owned nickel steel armour-plate factory."

The report created a deep impression on the House and Herr von Kardorff's attempts to defend the exposed profiteers proved futile. A resolution to establish a State-owned armour-plate works was adopted unopposed, except by a single vote! All the political parties of the Imperial Reichstag were united in their indignation over "the unprecedented profiteering" of the Essen firm.

It is difficult to appreciate the motives which caused the Secretary of State, Tirpitz, from whose quiver the fatal arrow emanated, to take the action he did. The fact that at the moment, as he himself revealed, he was fighting for a price reduction and found little co-operation from Essen, may explain some things. And when Krupp declared later that he had offered a voluntary reduction, that reduction was a very small one. Even after the passage of the Navy Acts, which ensured huge contracts for them, profiteering by Essen continued.

Armament business is, however, usually more complicated than disputes about a small percentage of increase or reduction in prices. After the Reichstag resolution the Cologne Catholic newspaper returned to the charge with the embittered statement that notwithstanding a recent reduction in prices, Krupp and Stumm were still making huge profits. It gave as proof of this statement the announcement that a "Rhenish syndicate" had recently declared their readiness to supply nickel steel of identical quality, from 1903 onwards, at a price of 1550 marks—that is, at 770 marks cheaper than Krupp—subject to an assurance that a corresponding allotment of future contracts would be guaranteed to them. The terms of this confidential offer had become known to the two privileged firms concerned, who thereupon reduced their price from 2320 to 1920 marks on condition that the whole of the contracts up to 1907 should be placed with them.

This undoubtedly puts a fresh complexion on the

peculiar case of *Tirpitz* v. *Krupp*. All circumstances point to a concealed manœuvre intended to take the wind out of the sails of the Rhenish syndicate, and of its chief instigator Thyssen, who had obviously inspired the Cologne press attacks. By publication of the details of the American dispute about armour prices (which were, of course, perfectly well known to their opponents), the German Admiralty succeeded in demonstrating their complete detachment from their compromised contractors. They could, moreover, in consideration of the latter's agreeing to reduce their prices, compensate them by undertaking to place all future contracts with Krupp up to 1907—completely ignoring the Reichstag resolution. The upshot of this first public accusation brought against the Essen firm's profiteering was a formal extension of their monopoly for a further five years—and a Parliament resting on its laurels!

It was two years before the next Krupp debate occurred. Germany's expenditure on armaments had risen to over 1000 millions and the machinations of the contractors were the main cause of public uneasiness on the subject. The Imperial Dockyard in Kiel had invited tenders for the supply of 410 tons of armour plate from eight manufacturers—including Krupp. Shortly afterwards the press published an appeal from the Steel Plate Manufacturers' Association (of Essen) to the firms concerned, inviting them to come to an agreement "to avoid price cutting." Each manufacturer was to raise his quotation by 30 marks per ton, and should the contract be secured, this excess was to be paid into a fund for the purpose of compensating the firms whose tenders had been rejected. Although the amount at issue was small, this system of organized price raising attracted attention.

The relations between Krupp and Stumm were also investigated and it became known that Krupp paid his Saarland ally a substantial compensation for the latter's agreeing to curtail his output of nickel steel.

The Budget Commission of 1903 now had its slogan: "Profiteering rings!" Secretary of State Tirpitz attempted to minimize the importance of the steel plate affair, although he knew quite well, even if he refrained from saying so, that this undertaking, in which Krupp played a leading part, had just compelled the great German shipping firms to agree to an increase of five marks per ton in the price of all ships built in the Empire as compared with that which obtained in Britain. The Minister of War was exposed to even sharper criticism over the cost of the ordnance supplied from Essen. The Krupp-Stumm arrangement found its counterpart in the Löwe-Mannlicher pact, the two last-named firms having mutually agreed that for every rifle supplied by one of the two firms the other should be paid compensation at the rate of two marks. Minister von Gossler, ever a welcome guest at Hügel, had perforce to admit that in the armament industry "competing firms are apt to come to agreements amongst themselves rendering it possible for them to quote very high prices." Curiously enough, nobody inquired about the progress made with the scheme for establishing a State-owned armour-plate works, approved two years earlier. However, the Commission once more made a recommendation, which was reported to the House: "To request the Imperial Chancellor to take every care that in the interests of national finances, free competition by the various contractors tendering for the supply of munitions and ordnance shall be maintained." A corresponding resolution was adopted by the Reichstag on March 13th, 1903, as recorded "by an overwhelming majority." It was the second formal censure on the Krupp profiteering monopoly.

How it happened that this conflict in the Reichstag now began to wane and finally died down completely without producing any results, is a matter which belongs in the Parliamentary history of Germany. Powerless in the face of the Government nominated by the Emperor, aimless in their resistance and superficial in their criticism, the Left Opposition could make no headway. Their complaints of the armament policy, more especially against Krupp, were invariably of an extempore nature, not followed up by any well-informed criticisms, and usually based on some casual indiscretion betrayed by a competitor—by Thyssen, Ehrhardt, or the Schichau shipyard—which would not exceed certain limits. Their horizon was correspondingly limited; the relations between big politics and big business, internal reaction and industrial agrarian centralization, were altogether outside their ken.

The futile issue of the debates of 1904 was deplorable. Although the activities of the Navy League, inspired by Krupp, which called for a shipbuilding programme of 300 millions and had earned a congratulatory telegram from the Emperor, coupled with the extremely costly introduction of the long recoil cylinder gun, provided ample grounds for a thorough clean up, the Budget Commission and the full assembly of the Reichstag wasted their time over questions of minor importance. The Catholic Centre Party denounced Krupp's practice of Sunday labour for the fulfilment of armament contracts, and Minister of War von Einem attempted to excuse the permit by denying all knowledge of the circumstances, adding that he would never have given his approval for Sunday labour. The leader of the Miners' Federation, Otto Hué, dealt more seriously with the vaunted working conditions prevailing in Essen. He quoted official returns: the number of accidents in the Ruhr district per 10,000 workers was 139, whereas in Essen it was 184. The percentage of invalids in Prussia was on an average 56, while in Krupp's works it was 66. The "welfare firm" paid lower rates than other firms and refused to introduce the eight-hour working day even in departments of heaviest work. There should be no misunderstanding regarding "Krupp's policy of presenting everything in the guise of a great social and political achievement, in order to cover up defects in other directions." The speaker was the author of a standard work on mining and an expert on the pre-war working conditions prevailing in the Essen works.

The contrast between questions of vital importance and those on which Parliament frittered its time away, reached a climax the following year. One event alone, which should have sufficed to spur the Opposition into a semblance of action, was the announcement by Tirpitz of the proposal to submit supplementary naval estimates in the autumn of 1905. It was the official announcement of an intense campaign by the Navy League and by the subsidized ordnance press, to break away from the great Navy Act. At this juncture a debate opened in the Reichstag on the relations between Krupp and Ehrhardt. The issue of a political campaign degenerated into a discussion of competitors.

Instigator was Representative Eickhoff who accused the German Ambassador in Constantinople, Baron Marschal von Bieberstein, of misusing his influence for the benefit of Krupp. The latest Turkish contract of sixty million francs had been bid for by Ehrhardt's firm, the Rheinmetall works, with a quotation seventeen millions below Krupp's bid, but under pressure from the Embassy the contract was nevertheless placed with Krupp. Rep. Beumer endeavoured to defend Krupp by declaring that the Turkish authorities had distrusted the sudden price reduction made by Ehrhardt; but the Government spokesmen were more candid. Secretary of State Richthofen thought it "perfectly natural that the older firm, better known abroad, should be recommended." The quarter from which the German diplomats received the relevant hint was betrayed by the Minister of War, who admitted that the Foreign Office had received instructions from the military authorities in regard to the matter. The fact that assistance of a non-official nature had been also resorted to by the firm of Krupp to smooth their path to the Porte was guardedly hinted at by their accuser Eickhoff inspired by Ehrhardt, who murmured, "I merely remind you of that ominous word—backsheesh."

The bickering between competitors again offered an unrivalled opportunity of extirpating the root of the whole evil, the private monopoly of the armament industry, on the occasion of the speech delivered by Minister of War von Einem in the Reichstag on March 27th, 1905: "Inner tubes for certain types of guns, which Krupp charged 35 marks for in 1897-98, were offered by Ehrhardt at 30.15 marks and the price quoted by all the firms was gradually reduced to 17.20 marks. Inner tubes for the model 1896, 8.2-inch guns were originally priced at 102 marks by Krupp and then at 89 marks by

Ehrhardt and eventually the price fell to 67.50 marks. I could quote other examples of similar price reductions."

Von Einem confined himself to these examples, because the more glaring ones were already common knowledge. It was well known that Krupp had to reduce the price of field guns from 4800 to 1900 marks, and of certain projectiles from 45 to 17 marks. The fact that even these reduced prices included a big margin of profit was sufficiently demonstrated in the case of the armour plate, where a price reduction of from 2320 to 1920 marks paled into insignificance beside the fact that the cost of production was only 900 marks.

In the face of such proofs of barefaced robbery of the State through private armament profiteers, it is incomprehensible that the Opposition should have allowed themselves to be misled by the question of Krupp versus Ehrhardt. Contemporary opinion might have considered the fight against the Essen monopoly to be at its fiercest stage, but we now know that what was possibly the last opportunity to deliver a deadly blow to it was let slip. The political helplessness of the German Parliament was almost excelled by its simple-mindedness and by its defective memory. Nobody remembered earlier resolutions, no one inquired about the State-owned armour-plate works, nor did anyone suggest similar measures for the manufacture of guns and munitions. Eight months later the Government closed the whole question by a supplementary naval bill.

All that remained after this miserable collapse of the campaign against the firm of Krupp was the business of the American armour-plate prices disclosed by Tirpitz and periodically resurrected as a "sensation." Tirpitz was, however, discretion itself compared with the French

Minister of Marine, Pelletan, who proclaimed, at about the same time: "There is a regular oligarchy of naval contractors, who for many years past have been robbing the State; it is supported by a man who plays a leading part in the administration. I am surrounded by traitors."

No Imperial Secretary of State for the Navy would have spoken thus, not even if Herr von Tirpitz's post had been held by one of the opposing Centre Party or by a member of the Progressive Party.

CONFLICT WITH LE CREUSOT

The armament industry had long since realized its own ideal of establishing the autarchy of arms, and it was now intensively wooing the half-hundred States who might become customers for supplies of guns, armour plate, projectiles, and explosives. The fight for this trade was a series of steel threads stretching through the major political issues of the decade preceding the Great War and was characterized by a bitter rivalry, Krupp versus Schneider or Schneider versus Krupp. Speaking shortly after the war, the Deputy Edouard Barthé stated in the Paris Chamber: "If one carefully examines past political events, it can be safely and correctly affirmed that the Krupp-Schneider rivalry influenced relations between Germany and France."

Once again the two competitors' interests collided in South America, as they had done thirty years earlier. Krupp's position was backed by the political and economic influence of the Empire and it was not for nothing that the Legation in Buenos Aires was occupied by Minister von Waldthausen, son of an old Essen family of millionaires. With the exception of Peru, which depended on Paris, the South American States at the beginning of the century constituted a Krupp monopoly which Le Creusot was out to attack.

In Brazil, when competitive tenders for the supply of quickfiring guns were opened in 1902, Schneider successfully opposed Krupp with a quotation for the supply of their "75" guns. The Brazilian Ordnance Board, however, was not quite satisfied about their performance and made arrangements for further tests in 1903. Schneider was again unable to make any progress; on the very morning of the day fixed for the firing tests, a destructive fire broke out in the warehouse in which the French firm's material was stored and all their guns were destroyed. Krupp agents declared that the French officer in charge started the fire himself to avoid defeat in the comparative firing tests—which in the light of past achievements in other countries sounds highly improbable. On the other hand, Schneider's representative accused the astute agents of the Essen firm of starting the fire. Be that as it may, a third series of firing tests was arranged to take place in February, 1904. The new guns arrived from Creusot at a Brazilian port in ample time for reloading them into river steamers, when an unexpected obstacle arose; the steamship company declined to transport them up-country because "dangerous" ammunition was included in the cargoes. When they finally gave way on March 22nd, the Brazilian Testing Commission declared that the period of grace had expired. But Schneider would not admit defeat and an offer to defray the cost of further demonstrations put the Brazilian authorities in a quandary. A refusal to accept might appear highly suspicious. And now appeared one of those agents so characteristic of the twilight proceedings of the armament world. Captain von Restoff of the Imperial German Artillery was also the South American representative of both Berlin and Essen. The press of Rio de Janeiro, and in particular the newspaper with which Captain von Restoff was connected, suddenly began an alarmist campaign to the effect that Peru was being armed by Schneider with a view to invading the Amazon Territory. It was a matter of life and death for Brazil to arm immediately and by a wholly fortuitous coincidence a complete artillery park was available in Essen, for immediate dispatch. Schneider's protest and assurances that Peru had not at that time ordered a single gun from Le Creusot were unheeded, and the final firing tests did not take place. Krupp secured the contract.

Such contempt for awkward rivals was not always possible. In the case of Argentina, for instance, where a complete re-equipment of the artillery with sixty batteries of modern quickfiring guns was envisaged and for which competitive tenders from Armstrong, Vickers, Schneider, Ehrhardt, and Krupp were invited, a formal assurance by the President himself had to be made that the verdict would be fair and genuine. Even so, Krupp obtained a preference; if his products proved equal to those of the other firms, his guns would be accepted in order to avoid a change of contractors.

On this occasion the backstairs influence of the two leading competitors seemed equally balanced. The Testing Commission's report of the trials carried out in the summer of 1908 stated that Schneider's products put up the best performance and that they therefore recommended both Krupp's and Schneider's tenders for acceptance. Such verdicts—which frequently are only a measure for the amount of "backsheesh" to be extorted —mean very little. The protests put forward by Schneider regarding concrete cases of favouritism shown to Krupp and the dishonest practices indulged in by their agents are more significant. Even if in the case of armament firms it is the pot calling the kettle black, there is an outside chance that the French may be believed in South America, as having been the victim of dishonest dealing on the part of their stronger opponents. Schneider vainly complained that Krupp was in the privileged position of being the only firm to give the Argentine gun crews preliminary instruction; that in the event of one of their guns bursting they were not, contrary to the usual practice, automatically disqualified, but permitted to exchange the damaged gun and were not called upon to submit to a breaking test. Schneider complained also that one of Krupp's fitters had been caught red-handed, tampering with the brake mechanism of a gun which had been officially sealed by the Argentine officers.

The Essen firm defended themselves by shrewdly delivered counter-charges. In order to frighten Schneider's friends in Parliament and on the Testing Commission, the Buenos Aires press was bombarded with a series of cables from Paris reporting serious gun accidents in the French navy, ostensibly due to the defective quality of the products of Le Creusot. This method of discrediting competition had been successfully resorted to by Krupp on more than one occasion. Minister von Waldthausen also did what he could, but the diplomatic and financial pressure brought to bear by Schneider proved strong enough to compel the Testing Commission to issue an amended report on the equal merits of both makers' guns. For all that, the Argentine Council of Ministers declined to permit the holding of the proposed final tests, as "taking too long," and the contract—worth millions—went to Essen.

Warned by the painful lessons of the last two competitive struggles, Krupp left no stone unturned in the third of the South American republics, Chile, in order to attain his object without complications. Although the Chilean Government had already invited Schneider to carry out a demonstration in April, 1909, it would still be possible to prevent it. In the midst of elaborate preparations a cable reached Creusot from Santiago curtly announcing cancellation of the open tenders. The contract had been placed with Krupp without any competitive bidding. The protests of Schneider and a claim for compensation at the rate of 200,000 francs were vigorously supported by the French Foreign Office, but Krupp's hand in Santiago was stronger and Chile did not even deign to reply.

These conflicts took place off-stage as far as Europe was concerned, but petty pin-pricks occurred nearer home, as, for instance, in the case of the tiny Republic of Andorra in the Pyrenees, where the single Krupp gun acquired by the State could not be fired without sending its projectile over the frontier into French territory. They have only a distant bearing on the change in Franco-German relations from cold courtesy at the beginning of the century to the mutual hostility engendered by the Entente Cordiale. The change was, to a large extent, brought about by events of the pre-war era which may be summarized in one word—Morocco. This was, at least as far as the first incident of 1905 is concerned, equivalent to Krupp.

Arthur Nicolson (later Lord Carnock), one of the men behind the scenes in the Foreign Office, has given a vivid picture of the grotesque scuffle at the Court of the Sultan of Morocco, where grooms, gardeners, musicians, slaters, and film cameramen contrived to win the confidence of the exalted ruler that they might further the advantages of the particular State to which they happened to belong. The German, a certain Herr Rottenburg, again occupied a responsible position; as a brother of the former Chief of the Imperial Chancellery, he was in close touch with the Foreign Office in Berlin, and as chief engineer, he represented the firm of Krupp. Morocco was of particular interest to Essen, not only as a dumping ground for obsolete arms, but because of the existence in Quenza of an extremely rich deposit of hæmatite, a valuable ore used in the production of ordnance steel. In contrast to South America the opponent in this territory (again Le Creusot) was stronger than Krupp. Allied with the financiers, who did not quibble about loans to Morocco, Schneider presented a strongly worded demand to the Sultan in the spring of 1905 to erect tariff walls round his country in order to exclude all imports other than French.

The events which followed constitute one of the dramatic episodes of the world's history and gave rise to much astonishment and heartburning amongst contemporaries. Hallgarten writes: "The Krupp representative Rottenburg, whose firm had more at stake in the matter than any other party, finally undertook to go to Berlin in order to warn Wilhelmstrasse of the proposed action by the French. He arrived just at the time when the firm of Krupp was worrying the Foreign Office (indignant with France but still irresolute) about the Bosphorus affair, *i.e.*, in the first half of 1905, when the great Mediterranean journey of the German Emperor was in course of preparation. What could be easier than to combine all these matters? This provides the key to the great puzzle and the reason for that fateful landing of the Emperor in Tangier on March 31st, 1905, behind which, then as now, "political" motives were imagined. The keen-eyed August Thyssen was the only one to see through the manœuvre. "The whole object of the move is to induce the Reichstag to approve credits for Krupp" was his resigned comment to the French journalist, Jules Huret.

Historians record that while still at sea, William II telegraphed to Berlin that he would not land. The private interests of the firm had in the meantime proved strong enough to induce the sovereign to act against his better judgment.

The political tension of 1905 was further strained by the situation on the Bosphorus, where the Bagdad Railway of the Deutsche Bank was being steadily pushed south, and consignments of war stores supplied by Krupp infused fresh vigour into the decaying state. Ever since the days of the Cannon King, the "Turkey" account had been a traditional item in Essen. On the other hand, the state of the "Essen" account in Turkey was looked upon by German diplomats as an indicator of the relations between Berlin and Stambul. Political action and business reaction frequently followed each other with suspicious promptness. Germany's brutal threat to blockade Greece during the Greco-Turkish war of 1897, was rewarded by the Porte with contracts for eight large armoured vessels which were placed with Krupp's Germania yard in Kiel. *The Times* (London) comment on the matter was that "it had been arranged directly between Yildiz-Kiosk and Berlin (*i.e.* between the Sultan and the German Emperor)."

In 1905 it was a question of equipping Turkey with recoil cylinder guns to the value of fifty million marks; this figure was high enough to influence even the most unyielding officials, and the gentlemen practising diplomacy on the Golden Horn were never noted for their inflexible rigidity. The German Ambassador, Marschall von Bieberstein, had already displayed activity in questions connected with the interests of great industrialists while at his former post of Secretary of State for Foreign Affairs. His French colleague and rival, Ernest Constans, was pressed by Delcassé to further the interests of Schneider, by doing his utmost to obstruct negotiations already commenced by Krupp. Unless at least one-third of the contract went to Le Creusot, Paris would boycott Turkish securities and close the Ottoman Bank. As France's investments in Turkey amounted to some 5000 millions she was not to be trifled with. Krupp's position was rendered difficult by being compelled to fight two opponents in different camps: Ehrhardt—whose defeat has already been referred to—and Schneider. "Backsheesh" assumed a multiplicity of forms, varying from open bribery of the Minister of War to the marriage of the two daughters of a high official to German noblemen—through the good offices of the cynical Marschall. Although it is difficult to prove the application of such practices to concrete cases, the fact that diplomatic intervention had been resorted to was here openly admitted. In their reply to Ehrhardt, the Essen firm said, "The support of the diplomatic representative in Constantinople must be considered as exceptional and due only to the unusually energetic action taken by the French Embassy on behalf of a French competitor."

That it was not "exceptional" had already been clearly demonstrated by Krupp in South America and it is equally improbable that the competitor was the first to seek such support. It is sufficient to note that official machinery on both sides was set in motion, as a cynical matter of course, for the furthering of commercial interests.

Krupp reaped most of the disputed harvest in the shape of contracts extending over a period of years. The Essen firm had formerly safeguarded themselves in similar cases by exacting deposits of scrip of future loans of equivalent value. On this occasion a financial syndicate was formed under the auspices of the Deutsche Bank, which in consideration of an advance of sixty million francs (most of which went straight to Essen) acquired a lien on the Turkish railway and tax revenues for a period of sixty-five years.

The extent of this double exploitation, by the bank and the armament contractor, can be appreciated only by examining what Turkey actually received in exchange. The more or less obsolete patterns of guns ordinarily sold to the Porte (indiscreetly revealed by a former Krupp representative) actually cost about 14,000 marks each to produce but were invoiced at 165,000 marks each. This accounts for much of the zeal displayed by the armament firms in wooing customers like Turkey, and for the large-scale corruption without which the acceptance by even the most ingenuous foreigner of such fantastic rates of profiteering would not be possible.

To revert to the Schneider-Krupp conflict. Monsieur Constans finally managed to save a fragment of the Turkish business for Le Creusot, but the start secured by the Essen firm on the Golden Horn could scarcely be made up. Painful friction in Morocco, the struggle just beginning in South America, and the Essen firm's activities in Serbia (which country was informed by Krupp that they were just as ready as Schneider to float foreign loans in consideration of contracts for the supply of ordnance), this underground war between the two armament titans overshadowed the relations between Germany and France in those years 1904-5, that are now considered to have marked the turning point in European politics. It was then that France decided to abandon her century-old antagonism to Britain and to conclude the *bonne entente* against Germany. Amongst the contributory causes to this decision the transactions of the Essen firm are not the least.

Up to August, 1914, Creusot exported twenty-two thousand guns, Krupp thirty thousand. Their sale was largely instrumental in splitting Europe into two opposing armed camps and in leading up to the events which caused them to be fired.

PRINCE CONSORT VON BOHLEN

Hügel castle was very quiet during these years. In other days one exotic flag after another had fluttered over it and those who were received at the Imperial Court hastened to pay their respects to the occupant of the residence in the Ruhr Valley. The business had become more impersonal since that time and the shadow of Capri still hung over the private life of the millionaire family. The austere *châtelaine* was not sought after for her hospitality.

The Emperor also had avoided Essen for four years. When he reappeared there in August, 1906, a shower of decorations descended on the firm. Mrs. Margarethe Krupp received the Order of William, while the Chairman and the Managing Director were awarded the Red Eagle. The Imperial visit was for the ostensible purpose of inspecting the Rheinhausen works, but it soon transpired that something special was in the air. Bertha Krupp, the heiress of the industrialist, was nearing her majority, and as a dutiful daughter she had just accepted the husband indicated by her mother and her solicitous Emperor.

Their choice had fallen on the thirty-six year old Councillor of Legation, Dr. Gustav von Bohlen und Halbach, a son of the Baden Envoy to Holland; born in The Hague, he came of a family with American blood in their veins. Both his parents were born in Philadelphia and his maternal grandfather, a General of the United States army, found a soldier's grave in Virginia. Young von Bohlen graduated at Heidelberg, attained the rank of lieutenant in the Baden forces and the title of Equerry at the Grand Ducal Court. After a short spell of legal work he entered the Diplomatic Service, became Secretary of Legation in Washington and in Pekin, where he came into touch with a sphere of particular interest to Krupp. Finally he was appointed Councillor of the German Legation at the Vatican. The local Essen press stated that, despite his youth, he was already the holder of British, Japanese, Chinese, and Austrian decorations. He was, in short, a model product of that system of Foreign Office training, which Bismarck termed "Die Ochsentour."

The Emperor came again to Hügel to attend the wedding on October 15th. He made a sentimental after-dinner speech in which he referred to "my dear Bertha," spoke of the deceased Krupp as "my valued and beloved friend," and concluded with effusion worthy of a travelling salesman, "May you be successful, my dear daughter, in maintaining the Works at the high standard of efficiency which they have attained and in continuing to supply our German Fatherland with offensive and defensive weapons of a quality and performance unapproached by those of any other nation." On the following day he permitted the young husband to change his name to Krupp von Bohlen und Halbach "to ensure at least an appearance of continuity of the Essen dynasty."

The choice of young von Bohlen as consort proved to be a fortunate one; the model diplomatic scholar revealed himself as model husband, model Krupp, and model subject. Although not a "strong man" like August Thyssen, nor a brilliant opportunist like Stinnes, nor even a man of outstanding social abilities and connections, the new chief and future ruler of the industrial kingdom of Essen was a thorough and reliable desk man. At the age of forty-three he was still described by the American journalist Fred Wile as "a man of scholarly appearance with a boyish modest manner." He paid serious attention to the cultivation of good relations between the family and the Works. Rumours of dissension between the young couple were energetically denied by notices distributed throughout all departments of the Works. In 1907 the birth of their first son—named Alfred after the Cannon King—was officially announced to the Board of Directors, who returned thanks on behalf of the Works employees and expressed the hope that the "blessings of Almighty God" would be bestowed on the new generation of Krupp.

The proud father had joined the Board in 1906, together with Vice-Admiral von Sack (retired), the confidant of the Emperor, already referred to. The capital of the Krupp Company had at the same time increased to 180 millions, making it second in size to the Deutsche Bank amongst commercial undertakings of the German Empire. The whole of the block of new shares to a value of twenty millions was, of course, allotted to the Krupp family.

The year 1906 saw the first original technical contribution to the armament world by Krupp in the shape of the first German submarine—which event was discreetly permitted to remain in the background. Earlier experiments had been made by the Germania Yard with a submersible boat of seventeen tons, the results of which had encouraged the Russian Government to place orders for three of these novel craft. Several years passed before the usually naval-minded Berlin authorities made a move, and it was only in 1905 that Krupp's were entrusted with the construction of a larger and improved submarine, for which purpose the Reichstag had already voted one and a half million marks. The launching of "U 1" revealed the potential value of this new naval arm. It was of particular value to Germany, in that the British Navy had made no start in that direction while the size of its fleet rendered it especially vulnerable to attacks by submarines. Up to 1912 the total number of German U-boats ordered from Krupp was, however, only nine!

This "failure" of the Berlin Admiralty has given rise to much perplexity amongst post-War historians, although the answer to the puzzle is contained in a few simple figures. After 1906 the cost of a battleship from the Germania shipyard was forty millions and that of a submarine two millions; the margin of profit on the former amounted to some five millions and on the latter perhaps one-twentieth of that sum. Krupp's reluctance to produce such goods was, therefore, easy to understand, as the zeal even of an armament contractor was limited and his business interests were the driving force of the fantastic naval policy of Germany. This explains why neither the Navy League nor the leading articles of the armament press made any demand for the new weapon. On the contrary, when after the French naval manœuvres of 1906 the Paris newspapers announced that the behaviour of the submarines had been so satisfactory as to indicate that ten would have sufficed to destroy the entire British Mediterranean Fleet, the *Rheinisch-Westfälische Zeitung*, the most alert of all the Krupp newspapers, questioned the correctness of the statement.

Another invention of that time, the airship of Count Zeppelin, shared the same fate. The Emperor ignored it consistently; Harden's *Zukunft* let the cat out of the bag: "Those 'above' do not want to have too much noise made by airship engines, for fear of drowning the agitation for the Navy." The enthusiasm aroused in the entire nation by the technical performance of the dirigible airship left the Essen firm quite cold, but they were quick enough to grasp how lucrative the idea of the airship might prove in the reverse sense—that of anti-aircraft defence. The official jubilee publication of the firm stated:

"When the success achieved by Count Zeppelin in 1906 indicated that dirigible airships would play an important part in future wars, the firm of Krupp thought it advisable to evolve guns with projectiles specially designed to combat this new weapon of war."

While all Germany triumphantly hailed the airship and even the army authorities were gradually realizing its potential uses for military purposes, the Essen firm were constructing anti-Zeppelin guns. They were extremely mobile and fired high explosive shell for setting fire to the airship. At the International Exhibition of Aviation held in Frankfort a/M. in 1909, Krupp demonstrated these guns to the ordnance experts of the world. It was only natural that they should be eagerly bought by Germany's enemies.

In the unceasingly critical atmosphere of these years the significance of such transactions could no longer be passed over. This also applied to the self-advertising welfare work, which the firm every now and again attempted to use as a screen for their activities in other directions. Thus, after the death of Fritz Krupp, a huge sum was alleged to have been assigned for aged workers, whereas it was later revealed that all that it meant was a book-keeping transaction relating to a remittance of rentals—long before granted. On the marriage of the young Miss Bertha the Employees' Insurance Fund was endowed, although the subscriptions of its members had in the past amply sufficed to meet its requirements. But the fund's capital was partially (to the extent of a total of some thirty millions) lent to the firm of Krupp at an interest rate of four per cent., although the firm earned ten per cent. on its money. The extent of the sacrifice involved by such "gifts" may be accurately estimated by the fact that the post-War inflation caused them to shrink to zero again—to the ultimate benefit of the donors. This, however, was a matter of the still distant future; for the moment the Essen firm were content to parade their sacrifices without inviting comments on their earnings and to present the Emperor with an opportunity to speak of young Mrs. Krupp as "an angel of mercy."

Wrangles over the Pensions Fund give a more thorough-going criticism of the Krupp "Welfare." Every year there was an ever-increasing number of law suits brought by aggrieved ex-employees who had forfeited their claim to a refund of their subscriptions. The plaintiffs obtained a legal opinion from the renowned Swiss legal authority, Professor Dr. Lothmar, who described the regulations governing the Krupp Pensions Fund as "a crying outrage on the universally accepted standards of equity." In reply to a question put in the Reichstag, Imperial Chancellor von Bethmann-Hollweg stated that the fund referred to "is no ordinary means of insurance" but was established "on account of a desire inspired by purely business, and, I may add, perfectly legitimate, reasons for the retention of a permanent nucleus of employees." The law courts appeared to have held other views on the legitimacy of this kind of welfare. Time and again the Krupp Pensions Fund was condemned to repay a portion of the subscriptions paid in by employees who had been dismissed and in consequence the illegal regulations had to be altered.

The decade immediately preceding the War was a period of unfettered expansion and development for the firm of Krupp. Borne up by the boom in the German iron and steel industry and protected against temporary setbacks, as in 1908 by the turnover assured through the Navy Acts and the ever-increasing land armaments, the giant of Essen represented the economic flower of pre-War Germany. The firm's development after 1906 was characterized by a continual search for fresh sources of supply of raw materials. Lack of ore deposits of their own forced Krupp's to compete for the exploitation of any new deposits elsewhere. The acquisition by purchase of Prince von Solm-Braunfels's mines on the Lahn, of the Bieber mines in Gelnhausen, and of others in Hesse and Waldeck, did not suffice; it became necessary to import further large supplies of ore from Lorraine and Spain and eventually from Sweden and Lapland. The question of coal supplies was equally acute, as, although located at the heart of the Ruhr territory, Krupp had neglected to acquire any collieries adjoining the principal Works; and, shut in by coal mines belonging to Thyssen, Stinnes, and the Cologne Mining Corporation, they were compelled to mine their own coal supplies from more remote parts of the country.

A systematic reconstruction of the low-built and obsolete workshops erected half a century before, was now undertaken at the main works. Huge modern buildings arose in their place and served to house the various production departments, such as the foundry, the spring shop, the gunmounting shop and gun factory, the fuse and projectile turneries, which could conveniently be located together. In 1907 a new building for the production of steel by an electro-metallurgical process was erected. Additional blast furnaces were also built at the

Friedrich-Alfred works and still another firing ground laid out at Essen—the fourth!

All this was normal expansion but there was little corresponding upward growth in inventive initiative. It was not entirely chance that Krupp's competitors Thyssen and Haniel forged ahead and disputed the supremacy of Essen, with consequent clashes over quota questions for the steel syndicate. What may be looked upon as typical of Krupp development during these years, was the fact that their greatest achievement was represented by a gigantic new block of office buildings, designated the "Chief Administrative Office." It adjoined the small "Original Family House," and its vast rectangular bulk, some 200 feet in height, towered above the surrounding sea of cupolas, chimneys, and factory buildings dotted amidst a network of railway lines. Outwardly the head office was built in the castellated style popularized by William II, but its internal arrangements were on the most modern lines. The Chairman's room in these headquarters of Germany's armament production was destined to be the scene of fateful events in German history and to witness developments which led down to the present day.

The guidance of the Krupp Board of Directors had by this time been assumed by a man who was fated to be the evil genius of Germany's destinies for the next twenty-five years, Privy Finance Councillor Dr. Alfred Hugenberg. He first came into contact with Essen while serving as a senior official in the Prussian Treasury Department under Finance Minister von Rheinbaben, who had been a devoted friend and visitor of the Krupp family since the days of Fritz Krupp. Rheinbaben had pointed out Hugenberg as an energetic and ambitious egotist, whose arrogant and inflexible policy in connection with the various Government committees on which he served marked him as an eminently suitable person to handle the affairs of the Essen armament monopoly.

Hugenberg succeeded Herr Rötger as Chairman of the Board of Directors of the Krupp Company in the autumn of 1909. His relations with the members of the Krupp family appear to have been characterized by civility rather than cordiality. There was obviously little in common between the aristocratic Krupps and the plebeian upstart whose forceful and aggressive manner formed such a contrast to their own. But the trend of events in the history of the firm had frequently endowed strong personalities of his type with more real power than that enjoyed by the actual proprietors of the business. Ever since the defection of Hans Jencke the lack of a representative head of the firm had constantly made itself felt. Hugenberg's position was rapidly consolidated and his activities were not confined merely to the works in Essen. He became director on the boards of the Gelsenkirchen and Deutsch-Luxemburg Mines, President of the Colliery Association and the Mining Corporation, both organizations of the most exclusive magnates in the Ruhr territory.

The new chief executive of Krupp rapidly became the outstanding figure in leading German industries at a time when they were on the eve of far-reaching developments. The limits of internal industrial expansion had been attained and the crisis of 1908 was a first warning which caused the financial magnates of Germany to look further afield. The pressure of German imperialism was raised by many degrees.

The developments may be summed up in the name

"Dreadnought" and had a direct bearing on certain figures which were dealt with in one of the Essen offices.

THE GREAT PANIC

The last real co-operation between Germany and Britain took place in the winter of 1902, in the Venezuela Affair, and concerned the collection of private debts—including those of Krupp. They related to the construction of railways by Krupp, which had been financed by the Diskontogesellschaft. German publications estimate the amount of the Venezuelan indebtedness to both companies at twelve millions, while a Conservative member of the Reichstag, who was an avowed enemy of banking interests, calculated that the usurious interest charged on the debt raised its total to fifty million francs, justified by the explanation that a portion of the bank loan had been placed abroad. An indignant Venezuelan diplomat declared "Shylock is not only a well-known Shakespeare character; he can also be found in Berlin." In all fairness it must be pointed out that Berlin was only acting as sheriff for Essen.

It is not surprising that the Venezuelan Government demurred at payment and attempted to liquidate the debt by State loan bonds of doubtful value. But Wilhelmstrasse was always ready to take up cudgels on behalf of Krupp and succeeded in persuading the British Government to act with Germany against the defaulting debtor. After the expiration of an ultimatum British and German warships blockaded the ports of Venezuela and compelled the surrender of her navy. The German gunboat "Panther" bombarded and destroyed the coastal forts. Under the pressure of such arguments Venezuela gave in and agreed to terms of payment. Krupp's millions were saved, while the substantial costs of the demonstration and the consequent anti-German feeling which it aroused in the United States were debited to the German nation. Subsequent requests for the employment of warships to enforce demands on foreign States were prudently declined by Imperial Chancellor Bülow with the declaration: "Our action was purely an exceptional measure."

These exceptional measures were, however, not exceptions where the interests of Krupp were concerned. Even before decisions were taken which led to a final break between Germany and Britain, the Russo-Japanese war of 1904-5 strained relations between Berlin and London, and once more it was the affairs of the Essen firm which became contributory cause. Although on this occasion the Hamburg-Amerika Line of Ballin swallowed the lion's share of war contracts from Russia and Löwe supplied numerous machine guns, it was the activity of Krupp's Germania shipyard which Britain, as the ally of Japan, watched with such irritation. The completion of torpedo craft and submarines for Russia under penalty clauses for delivery, constituted a flagrant breach of Germany's neutrality, of which Krupp's were fully aware. Attempts were made to conceal the destination of the vessels by shipping them to Lübeck by rail for conveyance to Helsingfors by the Finnish S.S. "Ægir." The boats had been taken to pieces and declared to be "pleasure yachts" to conceal their contraband character, but the press raised an outcry and the disconcerted Lübeck police had to confiscate the shipment, "as it might be considered a breach of neutrality."

The excessive zeal of the local police gave the Krupp shipment the character of a major political issue. The Imperial Chancellor had just given Japan a formal assurance that she could count on "the strict and undeviating neutrality" of Germany. Would the Government take appropriate action against Krupp? The backstairs influence of the Essen firm did not leave the issue in doubt; barely twenty-four hours after the arrest of the "Ægir's" cargo it was released. Acting on the assurance of an expert from the Admiralty, the Berlin authorities had decided that the boats, component parts of which bore inscriptions such as "magazine" and "torpedo room," were undoubtedly pleasure yachts. The press revealed the name of this imaginative expert—it was Admiral Barandon (retired), a former director of the firm of Krupp.

Further shipments to Russia passed off without incident. Krupp and Ballin made money and considered the fact that their actions had brought the Empire to the verge of war with Britain as a merely unavoidable side-issue. Together with the Diskontogesellschaft, Krupp formed the "Shantung Syndicate," whose attempts to secure monopoly concessions called forth further protests from London. When Spain placed an order for heavy fortress guns, which could only be intended for use against Gibraltar, the Essen firm promptly executed it. The effect in London of this and similar transactions was left to the Foreign Office to deal with. British reaction expressed in an ominous transfer of the main strength of the British Fleet from the Mediterranean, where it had been concentrated for a century, to the North Sea, might be a threat to Germany, but it did not impress Germany's "big business" interest in the slightest. They replied in the usual manner—by agitation for an accelerated programme of naval construction.

After 1905 an armament race began on both sides of the North Sea, enlivened by panicky rumours of projected German invasions, secret armaments, and possible British attacks. Outwardly it appeared to be a campaign by worried patriots, in the Deutscher Flottenverein and the British Navy League, agitating for improved types of ships, more effective armour, and better guns. The fact that their financial backers were the builders of these expensive articles, throws some light on the origin of this armament fever, without, however, exposing it entirely. Only a retrospective review of this chapter of contemporary history reveals the crazy struggle for supremacy in armaments, which led to an open rupture between Germany and Britain, as a game in which the players pass trump cards to one another.

The most important item in the construction of the new Dreadnoughts was the vast quantity of armour plate, amounting to about 4000 tons per ship and costing some eight million marks, after 1906. Reference has already been made to the sale of the foreign rights of the Krupp patents whereby the firm acquired a direct financial interest in the growth of foreign naval armaments. In order to exploit the original invention of the American Harvey for producing armour plate of exceptional hardness from an alloy of steel and nickel, the armament firms of the whole world had combined to form the Harvey United Company, Ltd. Established in 1901, it remained in being until the very eve of the Great War and during the thirteen years of its existence its directors included:

German—

Ludwig Klüpfel, Heinrich Vielhaber, and Emil Ehrensberger of the Krupp A.-G.

August Gathmann and Fritz Säftel of the Dillingen Iron Works.

The London representatives of the Deutsche Bank.

British—

John M. Falkner of Armstrong, Whitworth & Co., Ltd.;

Albert Vickers of Vickers Sons and Maxim, Ltd.;

Edward Ellis of John Brown & Co., Ltd.;

John A. Clark of Cammell Laird & Co., Ltd.;

Edward W. Richards, President of the Iron and Steel Institute;

William Beardmore of William Beardmore & Co., Ltd.

American—

Millard Hunsiker of the Carnegie Steel Company.

French—

Maurice Geny & Edouard Saladin of Schneider et Cie;

Leon Levy of the Chatillon Company;

C. F. M. Houdaille of the St. Chamod Company.

Italian—

Rafaele Bettini of the Societa degli Alti Forni Fondieri.

The entire international gang, outwardly in rivalry with one another, worked together in harmony on the board of the Harvey Company, Krupps' directors held 4731 shares, Dillingen 2731, and the Deutsche Bank 1350. Under the terms of the agreement on armour plate with Dillingen, and close relations with the Deutsche Bank, the votes corresponding to the holdings of the two last named may be counted as belonging to Krupp, whose voting strength (8812) was thereby little inferior to that of Schneider (9862).

The extent of the direct profit which accrued to Essen through this holding has not been disclosed. There were also the licence agreements under which the Harvey Company had acquired the right to work the Krupp patents on armour plate. They led to litigation before the U. S. Court of Claims, which in December, 1905, condemned the Harvey Company to pay Krupp royalties amounting to 650,000 dollars from May, 1903. This works out at one million marks per annum to Essen from only one of the foreign firms for a single article of armament production.

At the time of the formation of the Harvey Company, 1901, a second international armament cartel was established for dealing with the output of nickel steel, the "Steel Manufacturers Nickel Syndicate, Ltd." The original membership was confined to the British firms of Vickers, Armstrong, Beardmore, and Cammell Laird, but in 1903 they were joined by Schneider, Krupp, and Dillingen, and in 1905 by Terni and by the Austrian firm of Witkowitz. This cartel handled the output of nickel steel not covered by the Harvey armour-plate process and therefore included ordnance and projectiles. According to a statement by Deputy Barthé in the French Chamber, Krupp held 210 shares in this concern, the activities of which were to constitute one of the darkest chapters in the history of the World War.

The ring of armament interests which cut across national frontiers and inter-State feuds was completed by the "Chilworth Gunpowder Company, Ltd.," in which the firm of Armstrong Whitworth and the "United Rheinischen and Düneburger Powder Mills" appeared. In this case the connection with Essen remained concealed for a long time, as it had never been the policy of Krupp to disclose information about their participation in such undertakings. But in the course of the action *I. Wild* v. *Krupp* heard in the courts in London in October, 1914 (!) the extent of the Krupp interests in the Chilworth Company and in lesser British munitions firms was revealed. They were stated to amount to a million dollars, an estimate which would be quite in keeping with the profit of thirty millions earned annually by the Essen concern.

These international ramifications of the makers of armour plate, ordnance, and explosives, which have only recently been brought to light, were of course quite unknown to the general public at that time. The conspirators against the peace of the world worked under cover of a cloud into which only occasional beams of light penetrated, and they made a point of conducting publicity campaigns under the auspices of the various "Navy Leagues" and similar bodies, to bring grist to the mill of the secret financial backers of these organizations, by agitating for increased armaments and consequently —better business.

Outstanding among German agitators of the period was General Keim, a hard-shell Imperialist, who had succeeded Schweinburg in the virtual leadership of the German Navy League. When the Essen firm was shown to be badly compromised by the shady machinations of this League, he offered them his services in a letter which was subsequently made public by the Reichstag member Bernstein. In the Navy League Keim rendered invaluable service to his patron Krupp. He was primarily responsible for the publicity campaign which preceded the Navy Acts between the years 1905 and 1908 and he was largely instrumental in ensuring Bülow's victory in the "Hottentot Elections" which enabled German Imperialism to strengthen its backing. This victory also was made possible by the combined strength of the Krupp press which alarmed public opinion by sensational disclosures regarding the weakness of the German navy (the *Rheinisch-Westfälische Zeitung:* "Scrap the junk!") and led to a veritable panic. Finally Keim, supported by the young Saxon lawyer Stresemann, turned against "slack Tirpitz" who had failed to urge adoption of fantastic proposals to increase armaments. The campaign of the Krupp agents proved too strong for William II's Government, which discarded Tirpitz and replaced him by Admiral Köster. While Germany was being driven into paroxysms over naval construction, the British section of the international armament ring had not been idle. The attitude of Campbell-Bannerman's Liberal Cabinet brought the British "Navy League" into being. Backed by British armour-plate manufacturers, this body raised a storm of protests against the unpatriotic disarmament plans which, in view of Germany's activities, it described as "dangerous."

With such perfect team work the armament industry was bound to flourish. The building of the gigantic British "Dreadnought," the first of the modern floating fortresses, was countered by new Navy Acts in Berlin, under which the five billion programme of 1906 was increased by 800 millions and that of 1908 by a further billion, which was practically equivalent to four additional battleships in each year. The Reichstag, confused by Keim's propaganda, accepted these demands without demur and likewise offered no opposition to the fact that the cost of the individual ships had well-nigh been doubled. The unanimous denunciations of Krupp profiteering were a matter of long-forgotten Parliamentary independence which the heat of Imperialist charge had effectively melted.

Nevertheless, certain actions of the Krupp directorate began to attract attention in Britain. The fact that the purchase of the Germania shipyard by Krupp ten years before constituted the first symptom of the coming German navy era, had not been forgotten there. And again, it was easy to discern Essen's efforts to provide technical facilities for a naval construction that went far beyond published figures. A special loan was raised by Krupp for this particular purpose with, as the press pointed out, "no security offered by the borrowers or demanded by the lenders." Lord Lonsdale, speaking in the House of Commons, declared that Krupp received State assistance in raising the loan—which may be considered as a pretty plain reference to the purchase of Krupp bonds by the Emperor. The fifty millions thus secured by Krupp were used for extending and improving the Essen armour-plate rolling plant and the shipyard in Kiel, thereby enabling the German Government to attain their objective—reduction by nine months of the time required to build battleships.

All this served to arouse uneasiness in London, but it required oil poured on the smouldering resentment of the British Government and public to turn it into a real conflagration. This noble task was duly undertaken by

Krupp's friends Armstrong, Cammell Laird, Beardmore, and John Brown.

Their spokesman was Mr. Mulliner, chairman of their associated company, the Coventry Ordnance Works, Ltd., who proceeded to assail M.Ps., peers, generals, and admirals with information which differed in a marked degree from that published by other armament-mongers; it was obviously authentic and so meticulously accurate in regard to the activities of Krupp as to indicate that it was inspired from a particular quarter.

"Are the Admiralty aware," asked Mr. Mulliner in a letter addressed to them, "that Messrs. Krupp are incurring enormous expenditure at the present time, to enable them to produce heavy naval guns and mountings rapidly? We have latterly had a great deal of business in Germany and find that Krupp has provided all the great machine-tool manufacturers with sufficient work to keep them busy for the next year or two. This increase of facilities will enable them to attain an output greater than the total productive capacity of Great Britain."

This did not fail to produce its effect and the Asquith Government duly invited Mr. Mulliner to furnish full particulars to a special meeting of the Cabinet convened for the purpose. The representative of Krupp's English friends told a sensational story. According to him, Tirpitz was secretly circumventing the Navy Law and was having two battleships built which did not figure in the published returns. The keel plates had already been laid in two shipyards, one of them that of Krupp. The building capacity of Germany's yards was now fourteen battleships at a time, which would enable her to challenge British naval supremacy.

Mulliner's revelations produced dismay followed by intense alarm in the House of Commons and the entire nation. The country was seized with a real war panic and the Government was immediately authorized to construct a complete series of new Dreadnoughts, the contracts for which were assigned to Cammell Laird, Armstrong, and John Brown.

British critics have frequently and vehemently described the Mulliner revelations as exaggerated falsehoods. This is true enough, although the publication of pre-War official German documents has confirmed the fact that the circumventing of the Navy Law by Tirpitz was officially connived at. The fact that the extensions and improvements of Krupp's manufacturing facilities were intended to cope with an accelerated programme of naval construction no longer admits of dispute as it has been revealed in the official jubilee publication of the firm. But how did Mr. Mulliner get to hear of it? British writers suggest that it was through spies or through Krupp's themselves, and the significant remark of Mr. Mulliner "we have latterly had a great deal of business in Germany" appears to point to this latter alternative. The connection between the party undertaking the secret building of the ships, with those who disclosed the fact, is in any case a close one. They were associated in international companies, exchanged technical information, divided markets, determined quotas, and gave each other many a hint. It will never be possible to establish the dividing line between business interests and considerations of national defence secrets.

Only the final end of this dividing line can be discerned with a tragical clarity; while Krupp, Armstrong, and Dillingen, Cammell Laird, Köln-Rottweil, and John

Brown continued their successful co-operation, the political ways of Germany and Britain drew apart. The coming of the Dreadnought era had led to an armament race by both countries which made all attempts at agreement, continued until February, 1914, illusory shadow play devoid of any real basis. The attempts were foredoomed to failure because heavy industry, after the crisis of 1908, needed the millions spent on armaments to produce a boom in steel, and secured it by playing cynically into each other's hands.

"KORNWALZER"

The beginnings of Essen's secret influence on the machinery of state in Berlin dates from the early friendship of young Alfred Krupp and the Master of the Horse Krausnick through whose chief, Prince Anton of Hohenzollern, access to the circle of Bismarck and William I was effected. With their rise began the great days of the Essen works. Sixty years had elapsed and out of the friendship with the Master of the Horse had grown a close network of intimate threads binding Berlin and Essen together. Krupp's grasped the State in its tentacles as an octopus grasps its prey, and the Government was not only its best customer but made its paths straight in the world outside.

All endeavours to throw light on these misty relationships must, at best, be considered purely tentative. Definite proofs in substantiation of business transactions can only be found on the lower slopes of the oligarchic mountain of Krupp business. In its higher sphere conclusions can be drawn only from the personal contacts of those who dwell there. Following the heiress Bertha, the second daughter Barbara also chose her husband, Baron Tilo von Wilmowski, from the ranks of the higher State bureaucracy, thereby maintaining the character and objects of Krupp marriages. As son of the former head of the Imperial Chancellery, Baron Tilo brought valuable contacts to the firm.

The relations between Krupp's and the highly placed civil and military functionaries who adorned their administrative posts, naturally began long before their entry into the firm's service. The confidant of the Emperor, Vice-Admiral von Sack, already mentioned, had been Director of Contracts at the Berlin Admiralty and during the budget debate in the Reichstag showed himself a zealous partisan of the great industrial interests. After years of business dealings worth millions with Krupp, the Gontard Arms Factory, and the Powder Trust, he joined the boards of these concerns. Tirpitz's ingenuous assurances that he decided all questions of contracts himself and that von Sack merely advised him, may be regarded as a futile prevarication. Ex-Minister von Thielen and Admiral Barandon (retired) also held responsible executive posts in the Government, in which they came into close contact with Essen. The cases of all these gentlemen are covered by a statement made as witness in court by the former general manager of the Krupp office in Berlin, the retired artillery officer von Metzen: "In the latter years of my service I came into contact with Krupp's in connection with contract deliveries and joined their employ."

The duties assigned to these Krupp executives were invariably designed to make the most of their former official connections. The commercial director of the

"War Material Department" in Essen was General von Metzhausen—succinctly described by another Krupp director, Herr Eccius, as being "on the make." This same Eccius, described on the firm's list of directors as a "retired barrister," was actually a Councillor of Legation under the Foreign Office and therefore an eminently suitable person to take charge of the Foreign Contracts Department in Essen. When the Reichstag member Liebknecht gave evidence in court on matters concerning shady Krupp transactions abroad, "carried out through their Director Eccius," the presiding judge abruptly pulled him up. Successor to Eccius was another Foreign Office official, Dr. Wilhelm Mühlon, who did not prove of much value to the firm. Then there was another director, Captain Mouths, late of the Artillery, with excellent contacts with the military hierarchy dealing with the ordnance contracts which concerned the firm so intimately.

The exact method of "working" Berlin, practised by the frock-coated or uniformed agents of Krupp, is a secret chapter of which only a few isolated episodes can be uncovered, but even they suffice to provide an insight into "the methods other than business ones" so discreetly referred to by Ehrhardt. There was Captain Max Dreger, holder of a degree of Doctor of Engineering and formerly Inspector of Fire Arms at the Ministry of War, who spent several years in Essen until the firm realized how much greater the value of his services might prove to them if he were brought into close contact with his former colleagues. Therefore, according to his own subsequent statement, he "was instructed by Herr von Bohlen to take up his residence in Berlin and to assist the firm in their negotiations with the persons known to me." At this time—in 1912—there were already two Krupp representatives in Berlin with clearly defined duties; the Sales Manager Brandt, for private negotiation with junior officials, and Captain von Metzen, general manager for negotiations with the Ministries. What was Captain Dreger to do there? He was assigned to the highest spheres, to deal with his former colleagues and contemporaries who were then generals with considerable influence in matters affecting technical and personal questions. In their private capacity these powerful excellencies were very kindly old gentlemen, who would have hated to decline a favour. In the course of the subsequent Brandt trial, Dreger was charged with having been approached by Brandt on two occasions with a request that he should use his influence with the Ordnance Testing Committee to obtain an appointment there for a young Lieutenant Hoge, who could act as a reliable intelligence agent (or spy) for the firm. Dreger admitted that he was, on two occasions, approached with such a request, but denied that he passed it on to his friend General Bücking. But Hoge was in due course appointed for duty under the Ordnance Testing Committee, so that the coincidence may be regarded as a mysterious case of . . . telepathy.

In other respects the connections between Essen and this important committee, which had to approve all contracts and pass all contractors' deliveries on receipt, were by no means mysterious. Ever since the Cannon King brought about the fall of a hostile head of department, all doors were open to Krupp. Major Wangemann, one of the senior officers, was a regular agent of the firm, which had an agreement with him concerning supply of information. "We could obtain earlier advice about changes of personnel through him than through our Berlin office," admitted Director Eccius when on trial. After leaving the committee, Wangemann proved so useful to the firm that they allotted him a monthly salary. He was a zealous writer on military subjects and editor of a periodical relating to ordnance matters, but was cautioned by Eccius to conceal his employment by Krupp when "working" the press, to avoid being compromised.

The firm was anxious to see Wangemann's friendly relations with the officers of the Ordnance Testing Committee maintained, and resorted to a highly original means to this end. Wangemann was entrusted with the organization of a club of his former colleagues that he might be able conveniently to pump them and influence them. When on trial, Eccius attempted to repudiate the suggestion by explaining: "I merely heard of a voluntary assembly of the gentlemen on the Ordnance Testing Committee, in which Wangemann played a leading part."

Wangemann himself admitted that these assemblies took place regularly once a month, but that they were quite harmless. It is true that, year after year, the senior members of this official body either entered the service of Krupp themselves or placed their relatives in it (in 1913 the President's son received an appointment in Essen), but for all that the Wangemann Club must be regarded as a perfectly innocent association of honourable gentlemen.

However, the firm did not entirely rely on contacts with "heads." Even in such delicate matters they displayed typical German thoroughness in the distribution of their favours both to seniors and juniors. Dozens, often hundreds, of army people, officers and N.C.Os. were sent to Essen on duties connected with the testing and overseeing of ordnance work. These official overseers always found Krupp a veritable fairy godmother; although in receipt of full pay and allowances, they were wholly maintained by the firm and it is indisputably established that they were provided, free of all cost, with quarters, furniture (if already available they received an "allowance" in lieu), full subsistence for household, including maidservant, even flowers for the house, free rail tickets, and souvenirs of their visit on leaving. The total expenditure under these headings for eight junior representatives of the Admiralty was no less than 50,000 marks. As the service personnel frequently did duty in Essen for several years on end, the only result of such permanent sponging could be wholesale corruption. Krupp subsequently sought to defend the practice by claiming that the granting of "free board and lodging" was approved by the Berlin authorities, a statement which produced a hesitating denial on the part of Lieut.-General Wandel, brother of the Krupp lawyer of the same name: "The Army Council had no knowledge of any such approval." The force of public opinion eventually compelled the firm to abandon the system of free maintenance, but it was then replaced by direct allowances in cash. These involved an elaborately mysterious method of book-keeping, under which the "compensation" paid to an officer might appear to amount to only 160 marks, whereas it was 13,000 marks. The openly expressed disapproval of the Army Council under pressure of the Reichstag, had no effect on the corrupt system. Krupp's would not permit their kindly forethought for service overseers to be diminished, so, *honi soit qui mal y pense*, they merely booked the expenditure involved as "welfare work" in their accounts.

In the end it led to a first-rate scandal, the features of which are particularly dramatic under German conditions, the case of the sales manager Brandt—the "Kornwalzer" affair.

Maximilian Brandt was Krupp's Berlin sales manager. In the centre of the capital at No. 19 Voss-strasse, a plate bore the inscription: "Fried. Krupp A.-G. Materials of War." Former artillery officers, mercantile assistants, and agents were employed here in large offices and a special press department maintained touch with the Berlin press world. The general manager was Captain von Metzen, whose presence in Berlin was intermittent, as he spent most of his time selling arms in Belgium where, in his own words, he "enjoyed the special confidence of the Belgian War Minister." In his absence Brandt had charge of the office.

This one-time storekeeper of the Ordnance Testing Committee was a reliable hard-working ranker. He showed great discretion in the discharge of confidential missions while employed in Essen and was, by special resolution of the directors, transferred to Berlin. No one seemed to know just why, strangely enough. And none of the ingenuous directors appears to have seen anything incongruous in the fact that this junior employee's salary rose from 4000 to 8000 marks, and was again increased to 13,000 marks in 1912.

The activities of the quiet reserved man, who was official head of the agency, would probably never have come to public notice had it not been for an anonymous letter, enclosing seventeen mysterious slips bearing the heading "Kornwalzer," which was sent to Liebknecht, anti-militarist and radical Reichstag member. One look was sufficient to show that it concerned espionage reports from military establishments in Berlin, which, as the writer stated, had been collected by this Mr. Brandt for transmission to his directors in Essen. Liebknecht immediately sent a confidential report on the matter to War Minister von Heeringen.

The reputation of this distinguished Socialist, known throughout all Europe, probably caused the Minister to pay more attention to the communication than he might otherwise have done. The contents of the "Kornwalzer" reports were immediately recognized as confidential matter. Heeringen instituted a discreet censorship over the correspondence of the Krupp office in Berlin and within a few days he was in possession of further "Kornwalzer" papers. Police observation indicated that Brandt was in the habit of meeting persons in plain clothes who had been identified as subalterns of the Ordnance Testing Committee and Ordnance Store Department, and as employees of the War Ministry. A secret conference took place between the War Minister and the Director of Public Prosecutions who decided on immediate and simultaneous action in Berlin and Essen. In Berlin the following day the War Ministry secretary Pfeiffer, Lieutenants Hoge, Tilian, Hirst and Schleuder, the N.C.Os. Dröse and Schmidt, together with Brandt himself, were arrested. District Magistrate Wetzel searched the head office building in Essen and confiscated 700 espionage reports, signed "Kornwalzer," carefully filed in the safe of the former officer von Dewitz, who was also taken into custody.

The praiseworthy zeal displayed up to this point by the War Minister now suddenly slackened off. Although the confessions made by Brandt, shaken by his arrest, provided ample material for a complete exposure of the entire network of espionage and corruption spread in Berlin by the Krupp firm, no further serious inquiries were made into it after February 7th. The powerful friends of the threatened firm had come into action to quash the whole affair. Herr von Dewitz, the arrested officers, and finally Brandt himself, were released from custody. The matter threatened to peter out completely.

It required renewed vigorous action on the part of Liebknecht to press the charge against Krupp. On April 18th, 1913, he made a first report of the occurrence to an attentive and crowded Reichstag. Krupp stood accused of espionage for the purpose of "obtaining information regarding the contents of secret documents relating to designs, results of tests, and more especially prices quoted by or accepted from other firms."

The Minister of War could only confirm Liebknecht's statement and expressed his "unqualified disapproval of such a procedure." But he was careful to add that "there is, however, no evidence that the Essen directors were a party to it." Fearing the animosity which even this mild expression of official indignation might bring down on his head, the unhappy Heeringen concluded with a pæan of praise for the patriotic service rendered by the firm of Krupp.

The same evening the Chairman of the Board, Privy Councillor Hugenberg, arrived in Berlin and prepared a counter-blow. The Brandt creature was to be sacrificed to save the good name of greater men. These gentlemen could not possibly have seen anything suspicious in the contents of the Brandt espionage reports as "confidential reports are a matter of everyday occurrence with us."

Then Hugenberg uttered the cynical catch-word, which served as a cue to the press, the court, and the Government, to turn the scandal into a charge against the accuser: "There is no Krupp case, but only a Liebknecht case!"

First to take this attitude and once more afford the protection of his "Shield of Honour" to Krupp's in their hour of danger, was the Emperor. Two months after the accusation in the Reichstag, on June 19th, William II bestowed the Order of the Red Eagle on the head of the compromised firm, Herr Krupp von Bohlen und Halbach.

After this exhibition of unmistakable partisanship by the All Highest, any further legal proceedings in the case could be little more than a farce. Of the ten directors originally called to assist at the preliminary investigation (Hugenberg, Rötger, Haux, Rausenberger, Mouths, Dreger, Mühlon, Eccius, Marquardt, and Dewitz) only one—Eccius—was finally recognized as being connected with the case. Less regard was shown for Brandt, the subalterns, and the N.C.Os. who were all arraigned before the Court in what now appeared merely a trial of minor offenders.

The prosecution deferred further action. At the trials held in August for the military offenders and in October in the cases of Brandt and Eccius, the utmost care was taken to preserve secrecy on all vital issues, and the judges took care to shut off all dangerous side-issues. When the witness Rausenberger declared: "We can always obtain all the information we require" and Colonel Brandt confirmed this by saying "There are no military secrets where Krupp is concerned," the presiding judge discreetly passed on to other questions.

When von Metzen became more explicit and stated that the expression "underground contracts" was commonly used in the firm's business, the prosecuting counsel displayed no curiosity. No comments were made on repeated assertions by the accused army men that memoranda had been circulated among them, making preferential treatment for Krupp and agents obligatory. Great stress, however, was laid on "favours" shown by Brandt to the accused military personnel, when in bars or restaurants he pumped them for valuable information as to details of designs, costs, tests, etc. The efforts made by the court to minimize the importance of the whole case are quite obvious. And yet the trifling outlay involved in the treating and the petty sums that changed hands was a sign of clever work, as a lavish expenditure would infallibly have aroused the suspicions of the subordinate military personnel concerned.

Very few of the 700 secret reports originally impounded were publicly read, and then only after a rigorous censorship. They all related to quotations and designs of the firm of Ehrhardt and will be referred to later. The peculiar name "Kornwalzer" was stated by witnesses to have been selected at random from a telegraphic code.

Examination of the Krupp directors led to an orgy of unabashed lying. An original letter from Marquardt to von Metzen, produced in court, contained a reply to a query from the latter, to the effect that the "Kornwalzers" were now in the hands of all executives. And most of the reports were countersigned by those for whom they were intended. Nevertheless, they all swore they had read nothing, or if they had, had thought nothing of it. This parade of innocence caused even the prosecuting counsel to deliver the following sarcastic summary:—

"In the course of six and a half years, Brandt sent approximately fifteen hundred reports to Essen. The reports give a clear insight into the whole of the working of the War Ministry, the Ordnance Testing Committee and the Ordnance Store Department, and they furnish the firm of Krupp with comprehensive information regarding the activities of all producers of ordnance material and of the prices quoted by their competitors. Nevertheless, if the witnesses are to be believed, nobody in Essen was interested in this information. Mr. von Dewitz circulated it among various people, but none of them read it. These highly placed gentlemen were much too busy to bother about such tiresome stuff. They merely affixed their signatures to it and if Mr. von Dewitz is asked what he did with the documents he tells us that after circulating them, he put them away in a cupboard."

Impervious denials came from the leading figure in Essen, Privy Councillor Hugenberg. He disclaimed all knowledge of Brandt's doings, although his co-director Mühlon had sought to defend himself by stating that he had reported to the Board of Directors. Hugenberg's assertion that the Krupp board meetings were not minuted caused a sensation. The secret negotiations with the compromising witness von Metzen "had been carried on by an independent lawyer named von Simson." The sworn evidence of Hugenberg omitted to mention that Privy Councillor August von Simson happened to be the Chairman of the firm's Advisory Council.

The only witness who attempted to show that the misdeeds of the minor offenders were negligible compared to those committed by the firm itself, was Liebknecht. He offered to produce further evidence of the relations between Krupp and the War Ministry and of the firm's intelligence system and its activities, as well as the corruption by them of foreign newspapers, and the shady transactions of the accused Director Eccius. The presiding judge curtly declined to hear it as being "irrelevant to the case."

The verdict was in keeping with the entire proceeding by its severity towards subordinates; the subalterns and N.C.Os. were sentenced to six months in prison and Brandt four months. Director Eccius, drawing a salary of some hundred thousands marks per annum, was punished by a fine of 1200 marks. No appeals were entered.

Apart from the trial the Brandt case led to the resignation of the Minister of War, von Heeringen, whom the Party of the Right would not forgive for his "pact" with Liebknecht. The Director of Public Prosecutions graciously accepted an humble apology from the pro-Krupp newspaper *Lokalanzeiger* for its reference to him: "Think of your career!"

Witness for the prosecution von Metzen and his Essen counsel, Dr. Bell (later one of the signatories of the Versailles Peace Treaty), were enjoined to silence by threats of Court of Honour proceedings against them.

Representative Liebknecht, nominated by his party to serve on the Reichstag Committee of Inquiry into armament contracts, was publicly rejected by the Government. One of the Krupp directors challenged him to a duel and the national press threatened him with violence.

Brandt's successor was Major Steinmetz (retired).

Did not Herr Hugenberg proclaim: "There is no Krupp case, but only a Liebknecht case"?

THE DOWNFALL OF EHRHARDT

The firm passed the years 1912-13 in the glare of publicity. There were jubilee celebrations marked by publications running into thousands of pages, speeches by the Emperor, scandals, and scandalous trials, weeks of debate in the Reichstag, and showers of press polemics. All these things had descended on Krupp of Essen, and it would almost appear as if the doings of the armament-mongers had now definitely been brought under the control of public opinion. Appearances are deceptive: the most important issue had, after all, remained concealed.

It concerned the Krupp monopoly. Agreements concluded between Berlin and Essen, which were renewed for a further period of five years in 1902 and again in 1907, were due to expire shortly. It will be remembered that they guaranteed an annual profit to the firm which even ten years before, and for armour-plate alone, had run up to six millions. Whenever the agreement neared its conclusion or renewal, competitors redoubled their efforts to secure a share of the business. This is illustrated by secret correspondence unearthed by the historian Hallgarten from Admiralty archives. It includes a letter, dated June 16th, 1908, from Secretary of State Alfred von Tirpitz to the owner of the Schichau shipyard, Privy Councillor Ziese, in which the writer expresses concern at rumours relating to the construction costs of battleships, circulated amongst Reichstag members after a visit paid by a Parliamentary delegation to the Yards in Danzig. Councillor Ziese is alleged to have made certain compromising statements regarding these costs. The writer is anxious to ascertain the exact nature of these statements in order that he may be in a position to deal with the questions he expects to have put to him in the Reichstag in connection with the matter.

Herr Ziese's reply, dated three days later, explains with ill-concealed sarcasm that he commented to the delegates on the fact that the manufacture of armour plate in Germany was solely in the hands of Krupp and of their cartel associates Dillingen, and that he himself would readily undertake the production of armour plate and ordnance were it not for the opposition he would encounter. Although there were five or six firms manufacturing ordnance, and four or five producing armour plate in Great Britain, it appeared improbable that the old-established German monopoly could be broken. Herr Ziese goes on to say "when the gentlemen asked me why battleships cost so much less to build in Britain than in Germany, I stated that this was largely due to the cost of the guns and armour."

Yet this Ziese incident was only an episode. The main attack on Krupp came from another quarter. The first bid for a share in the heavy ordnance contracts was made by Krupp's old rival in Dusseldorf, the Rhenish metal works of Ehrhardt. A potential competitor in armour-plate production was August Thyssen, steel magnate of the Ruhr, dangerous both from the vigorous policy he pursued and the successful technical enterprise he displayed. For many years past Ehrhardt and Thyssen had opposed the Krupp monopoly and the question now arose as to whether they were going to be able to break it.

The year was 1913 and not 1901. This meant more than mere difference in time. Imperialism had finally destroyed the always fragile structure of Liberalism in Germany. The former inconveniently "public" forum of debates on armament disputes had now been eliminated, *i.e.*, the annual public report of the Reichstag Budget Committee, in the course of which the Government was forced to supply inconveniently complete information on Army and Navy Estimates. The fact that no minutes were kept of these debates proved an insufficient safeguard, so on the pretext of "numerous complaints from manufacturers" (as the War Minister admitted) the Imperial Government proceeded to impose a preliminary censorship on this last remnant of Parliamentary control; critics of the Left were muzzled by special obligations concerning "confidence." Future conduct of the debates is typified by the following report of the Reichstag proceedings of February 17th, 1911—

"The Army Council announced that particulars of prices paid by the Government for war stores (guns, munitions, &c.) could be communicated only if they were treated as confidential. The National Liberals, Conservatives, and some of the Centre Party thereupon adopted a resolution to treat such information from the Government as confidential. But when the prices paid became known to them the members expressed their great regret that the impossibility of public discussion prevented taxpayers from receiving the interesting information that prices charged by private industry were considerably higher than production costs in the State-owned establishments."

Amusing, but very mild. In this case it was not the usual profiteer who was aimed at, as "private industry" comprised the privileged firms of Krupp, Dillingen, Löwe, Gontard, Koln-Rottweil, and a few others. A motion by the member Erzberger, a director of Thyssen, was more significant; he moved "that contracts also be placed with firms prepared to fulfill them at prices below those now being charged." This was aimed at the Krupp monopoly!

The Essen firm had prepared for this possibility years before. The measures employed were most meticulous and the utmost care had been taken to ensure that no trace of any of the intrigues in which they were involved could come to light. The fact that the French publicist Jules Huret allowed his book *En Allemagne* to discuss awkward subjects and to quote various witnesses hostile to the firm's interests could not, of course, be helped; but, curiously enough, a German edition of the book was published in which the passages offensive to Krupp were either toned down to incomprehensibility or were omitted entirely. One such passage concerned a statement by the usually silent August Thyssen—

"Poor Ehrhardt, who for the past fifteen years has been fighting with unparalleled vigour to get his products adopted! They have tried to ruin him; they have done all they could to discourage him and head him off. For instance, the Ehrhardt exhibits at the Düsseldorf Exposition were exceptionally fine and calculated to overcome the strongest misgivings. The Emperor came and spent half an hour on the Krupp stand, but never set foot on Ehrhardt's. Result: the guns will cost more, and will soon have to be replaced by new ones. How can Ehrhardt combat nepotism? Krupp's employees include two brothers of Cabinet Ministers, and the brother of the head of the German Navy."

This brings us back to an old theme, the fight of the monopolist against the outsider, big capital against a weaker, if technically successful, competitor. A generation before, the opponent was Gruson, now it was Ehrhardt. The result of the first encounter has already been related—Ehrhardt's guns proved to be the better, but Krupp secured the larger contracts. The overwhelmingly greater resources of the Krupps enabled them to represent their unfortunate competitors to the whole world as untrustworthy in business. They continually accused them of "illegally utilizing" Krupp patents, of "misleading publicity," and generally indulged in a campaign which did not scruple to adopt any means to achieve its purpose.

It required the Brandt case to dissipate the fog of business morality in which the Essen firm sought to envelop their own doings. The accused soldiers admitted that designs and prices of Ehrhardt were matters of special interest to the Krupp agents. The evidence of the accused Lieutenant Hoge was particularly disastrous to the honest indignation expressed by the Essen firm over the appropriation by Ehrhardt of the Haussner invention.

Presiding Judge: "In one of your 'Kornwalzers' to Brandt, you give full details of an experimental gun built by the Rheinische Metallwarenfabrik."

Hoge: "He only wanted to know how the gun differed from the Krupp design."

The Essen thirst for information was limited "only" to Ehrhardt designs. The special character of the armament industry renders it important to know the exact results of competitors' firing tests and the opinions of inspecting officers concerned.

Counsel for the Prosecution: "The important question underlying the 'Kornwalzer' report of the Artillery Test Committee findings concerns the industrial conflict between Krupp and Ehrhardt. If I am rightly informed Krupp's position in this conflict was not exactly favourable."

Expert Witness, Captain Ellertz: "Not at that stage of it."

Counsel for the Prosecution: "So it would have been in the interests of Krupp to secure advance information on the conclusions reached by the Artillery Test Committee?"

Expert Witness: "Yes."

The Essen intelligence service penetrated the most confidential transactions. For a firm having direct access and influence through a variety of channels to persons in responsible positions of authority, the following piece of information—taken from one of the few "Kornwalzers" read in open court—must have been of great value:—

"The Ministry of War has the transfer of the contract for the supply of tubular brakes to the Rheinische Metallwarenfabrik under consideration, to save payment of royalties. No final decision has yet been made."

No wonder Ehrhardt invariably ran up against mysterious obstacles in any negotiations, which always seemed to hang fire. Competitors secretly watched his designs, firing test results, and contract terms, ready to intervene at the appropriate moment with better proposals, intended to discomfit all rivals. Ample resources and brilliant contacts were allied to utterly unscrupulous methods, which, in accordance with the old Krupp practice, were carefully calculated to prove their opponent's undoing.

This policy of isolating competitors was not limited to Germany. The Russian publicist, Leonidoff, who had made substantial contributions to "The Policy of Armament Capital" from the Tsarist archives, mentions a secret agreement between Essen and Le Creusot:—

"The French concern undertook to support Krupp against other German competitors; should the German firm of Ehrhardt submit a lower tender than Krupp, Creusot were to underbid Ehrhardt and accept the contract, even if it involved them in a loss."

It was, therefore, useless for Ehrhardt to cut prices, as in the case of Constantinople, below production costs, because wealthier rivals could outlast him.

The pitiless battle waged by the monopolists received effective support from German officialdom. The heads of the army and navy consistently ignored the quotations of the Dusseldorf firm. "Where are we to get our ordnance? We have nobody except Krupp in Germany," Tirpitz proclaimed in the Reichstag, although he was perfectly well aware of the efforts being made by Ehrhardt to extend the range of his products beyond field artillery. This same Tirpitz ordered one (!) naval gun in Düsseldorf in order to be able to comment on the fully expected non-delivery: "We are still waiting for it." "The State is the same as Krupp," said the young lieutenants, to comfort themselves when they were caught in the snare of corruption set by Brandt. Ministers and Secretaries of State reacted in a similar manner. Consideration for the friendship of the Emperor with the millionaire family, common financial bonds between State and contractors, not to mention bureaucratic indolence which found a convenient monopoly easier to deal with than being harassed by competitive tenders, all played their part. In the great debate over the freezing out of Ehrhardt from Turkish contracts, Secretary Richthofen of the Foreign Office finally advised the Dusseldorf complainants to make common cause with Krupp.

An ominous suggestion! Inspired by the fate of Gruson, rumours of pending interests of Essen in the Rheinmetallwerk were always being circulated. Ehrhardt does not appear to have taken them seriously, and they were, on various occasions, denied by Krupp. So long as there was a majority hostile to the Krupp monopoly in the Reichstag, the firm's directors preferred a process of gradual throttling to any dramatic action and they had every reason to rest content with the result. Their incessant litigation eventually involved Ehrhardt in financial difficulties which hampered the growth of the works. The pressure of Krupp on the potential market for ordnance was so great as to reduce the credit facilities available for Ehrhardt, and on the eve of the great armament increase of 1913, the latter were desperately short of funds. To provide a substantial basis for negotiating a loan, Ehrhardt's had to extend their range of products to include guns of the largest size. The fatal pressure of Krupp compelled them to make a desperate attempt on the former's monopoly of heavy ordnance.

Here was the moment for the Essen firm to tighten the noose. At the general meeting of the Rheinmetall A.-G., held on August 20th, 1913, proposals for reconstruction were submitted which in the favourable circumstances presented by the general increase of armaments throughout Europe, appeared to have every prospect of success. But a show of hands brought a great surprise; a qualified minority representing three and a half million shares blocked the resolution. This block of shares was registered as being held by strangers, a gentleman of independent means, a broker, and a solicitor, who were obviously straw men. When the *Frankfurter Zeitung* asked if these people were nominees of Krupp, the answer was quite plain—there was no answer!

Krupp was now in the enemy's fort and although not yet in possession of a majority of the shares, could effectively block all the competitor's business schemes. The short-lived dream of breaking the Essen monopoly for heavy ordnance was over, Berlin now really had only one contractor to deal with.

There remained August Thyssen. He was tougher, more farsighted and able than the somewhat haphazard Ehrhardt, and within a single generation he had succeeded in turning blast furnaces, rolling mills, collieries, and hardware businesses into one gigantic and comprehensive trust, which was a model of an up-to-date rationalized industrial undertaking. Since his bold move in connection with the iron-ore deposits in Northern France, the squat little man from Mülheim had become an almost legendary figure. Krupp could not afford to concede the slightest advantage to an opponent of such calibre.

Notwithstanding the vigorous publicity campaign conducted by his Parliamentary agent Erzberger, Thyssen failed to make further progress. The opposition offered by the ministerial bureaucracy was surprisingly stiff and wrecked the negotiations for participation in the armour contracts in their early stage. It was useless for Erzberger to state in the Reichstag that ordnance and armour plate were being purchased from Krupp at prices forty per cent. above those quoted by competitors; the days of Old Prussian thrift were past now. Although definite proof, such as was available in 1902, was hard to produce, the hand of Essen can be clearly recognized in these intrigues. According to a statement made by Thyssen to Huret, the Government's reply to his inquiry as to the extent and nature of their armour-plate requirements, was "Ask Krupp." At first it seemed that the Parliamentary support given the Mülheim man by the Catholic Party was powerful enough to compel Tirpitz to give way. He agreed to place orders, but expressed his inability to enter into long-term contracts or to ensure continuous acceptation of the Mülheim processes. As the cost of equipping an armour-plate rolling plant ran into some thirty millions, Thyssen declined to undertake such a risk. Did the vague promise of the Secretary of State constitute a trap? He could scarcely hope to catch the shrewd and calculating Thyssen in it, but the warning served its purpose in appraising him of the strength of the Krupp monopoly guarded by a phalanx of imperial bureaucracy. He resigned himself to the abandonment of his armour-plate schemes.

The last serious competitor was thus held off and the renewal of the contracts for deliveries due in 1912 and 1913 was granted promptly. The Catholic Count von Oppersdorf states that they were renewed for even longer periods than before and on far less favourable terms—for the State of course.

THE DUEL OF THE GIANTS

Let us look back. In the course of our lengthy review of the history of the firm, we have repeatedly recognized the nature of the relations existing between high politics and big business. We have also seen how family friendships could be turned to good account in ordnance contracts, how alliances were concluded as a sequel to spadework by agents of armament firms, and how collisions between these firms could give rise to deadly enmities. Steel frequently cuts right across a political course and contrary to all accepted principles, finds its way straight into the camp of the enemy. Sometimes guns were just sold to parties who could pay for them, irrespective of political differences.

We must recognize that all that sort of thing was by now becoming hopelessly involved. The coming great cataclysm was not simply the result of the growth of opposing forces. The growth was general; rivalries and contacts, business transactions in terms of relativity, and obligations which business transactions imposed on firms effecting them. Difficult indeed to keep them apart.

The backdrop of the entire stage setting to the last pre-War period was provided by one definite development, dating from after 1906 and not entirely completed by August, 1914: the rearmament of defeated and disordered Russia. St. Petersburg was in need of everything —of an entire navy, artillery parks, fortifications, hundreds of thousands of rifles, even of factories and dockyards. The Tsar's Government put up 1300 million rubles (equal to about 650 million gold dollars) as a first installment and invited the international armament industry to apportion it out amongst them. Krupp came in too. The fact that every gun sold to Russia helped to close the gap made in the East in the encirclement of Germany was not among the matters considered by the Essen Head Office. Were Schneider and Vickers to be allowed a free hand? If the Emperor himself made it his business to try for such contracts, it became the duty of patriots to earn all they could on the job. Heavily laden trains, full of guns and armour plate for forts, left the works for St. Petersburg, and the Germania shipyard worked at high pressure on replacements for the Baltic Fleet sunk in Far Eastern waters. In the words of a Chamber of Commerce report, the works were "entrusted with so many orders, that it would keep them busier than ever before."

As long as the orders came of their own volition the parties interested in the Russian armaments business maintained an armed truce with each other; in fact they elaborated discreet working arrangements and contacted key positions of common interest, not, of course, in the crude manner of the cartel agreements which were referred to in connection with the Harvey company. On this occasion the gentlemen became quite intimate, although it was understood that each one had to look after himself; while they were shaking each other's right hand, their left hands were feeling for their neighbours' wallets.

The second Morocco conflict in 1911 was quite in keeping with the international interests of the various capitalist groups. Krupp, who had recently persuaded the Emperor to land in Tangier, had discontinued protests against the French policy of annexation and had gone over to the hereditary enemy's camp. The equally patriotic Schneider-Creusot found the German assistance most opportune, as their position in North Africa was being threatened by the "Société Française," a syndicate of powerful Paris banks, who had their eye on the valuable deposits of hæmatite ore. A rival syndicate was rapidly created under the name of the "Société d'Etudes de Quenza," with which the following firms were associated:—

Schneider-Creusot S.A.
Krupp A.-G., Essen.
Gelsenkirchner Bergwerke A.-G.
Guest, Keen & Nettlefolds, Ltd., London.
Cammell Laird & Co., Ltd., London.
John Cockerill S.A., Brussels.

German, French, British, and Belgian armament capital combined to go on with the good work. As the presiding geniuses of this international syndicate, the patriots of Le Creusot forced their French competitors to their knees and into amalgamation.

But the main attack on the interests of the Moroccan group who consolidated their position round the "Union des Mines Marocaines," came from beyond the Rhine, where the firm of Mannesmann, on the strength of dubious concession agreements, laid claim to the whole of the coveted ore deposits. These "German" claimants were also associated with French financiers, who helped to justify their onslaught on other German interests in Morocco, with which a certain patriotic firm in Essen was associated. To make confusion worse confounded there appeared in the Banque Union Parisienne, closely associated with the Krupp-Schneider-Morocco syndicate, a certain Mr. Basil Zaharoff, representing the firm of Vickers, to whom new deposits of ore were quite naturally a matter of interest. Thus this Franco-German conflict was characterized by an international conglomeration of armament firms on either side, who, as Jean Jaurès stated it in the French Chamber, "combine to secure the ore required for the manufacture of shells destined for their mutual destruction."

This was, therefore, one of the really great affairs of the armament industry, and not unnaturally the common task called for the time-honoured methods of inciting national hatreds as a means of furthering it. The armament press, led by the Essen *Rheinisch-Westfälische Zeitung*, seized on the "incident" of the German gunboat "Panther's" visit to the Moroccan port of Agadir as a pretext for arousing a war spirit, which came perilously near to explosion; subsequently Tirpitz admitted that on both sides certain precautionary measures were taken. It cannot be said that the speculation indulged in by the arms dealers of Essen, Le Creusot, and Sheffield was devoid of success; the German Government, for example, utilized the feeling of panic called forth by the Agadir incident to secure the unopposed acceptance of supplementary estimates for the navy, thereby bringing further profits to the firm of Krupp. A by-product, so to speak, of the profits from the Morocco deal coming to them via Paris. A third sequel to the whole business was reported in the press in the autumn of the same year: "The firm of Mannesmann entered into negotiations with Krupp to define and mutually protect their respective spheres of interests. The negotiation proved successful and Mannesmann have undertaken to buy 145,000 tons of raw steel per annum from

Essen. Herr von Bodenhausen, a Director of the Krupp A.-G., has joined the Board of the Mannesmann Works." Morocco was fertile soil!

The contacts between Essen and Le Creusot were more delicate in a territory like Austria, where the close proximity of the Balkans called for smart work. The greatest Austrian armament firm, the Skoda Werke in Pilsen, founded in 1859, and having a share capital of thirty-nine millions, passed through a financial crisis in 1906, and was reconstructed with the assistance of a group of Vienna banks headed by the Kreditanstalt. Since that time Krupp had financial influence in Austria, while Schneider, through their interests in the Paris group of financiers connected with the Austrian banks, had a similar one, even if the nationalist comedy prevented its disclosure. Relations with Essen gradually touch on the form of a cartel, and an agreement for the exchange of important patents was made between Krupp and Skoda. Incidentally, the latter provided a convenient screen for operations in China, where Yuan Shi Kai, in consideration of a loan from the Austrian Development Corporation, placed contracts to the value of some seventy millions, part of which were for guns of Krupp manufacture.

The urge for expansion felt by Krupp was accelerated by the sinister atmosphere prevailing in a neighbouring country, and it found an outlet in 1912 when the Hungarian Government considered the building of an arms factory for supplying the needs of their section of the Austro-Hungarian Army. Having got wind of this scheme before their competitors—including those in Pilsen—Krupps secured an option on the business from intermediaries in Budapest, and then proceeded to consolidate their position in regard to this somewhat nebulous proposition by inviting Skoda to participate in it for a consideration. In order to force the hand of the Hungarian Government Krupp then concluded a working agreement with the munitions factory of Manfred Weiss, under which the latter undertook not to cede any patents to a State-owned factory should the Government wish to establish one. When the works were finally established in Raab, Krupp took up only two out of thirteen million shares, but certain other features of the deal produced an unpleasant surprise when they became known. Krupp were to receive a cash payment of one million crowns and an annual royalty of 200,000 crowns for the use of their patents and processes. Certain contracts were also placed with Krupp "in accordance with the quotations they have furnished" as the text of the agreement had it.

It was a succulent morsel which Krupp had contrived to filch from the communal meal of their associates in Pilsen and the latter's secret allies, Schneider. But little matters of this kind did not upset the harmonious relations between gentlemen of honour, more especially as the opportune outbreak of war in the Balkans provided ample opportunity for revenge. At the outset Krupp's position in regard to the sanguinary conflict in 1912-13 was unfavourable. Schneider had forced them out of Greece after the dramatic armament clash in 1907 and the personal financial interests of King Ferdinand in Le Creusot involved the victorious Bulgarians in such dependence on them that even their own Sobranje began to kick. Krupp was, however, the chief contractor to the power which was forced to accept defeat after defeat—Turkey. The publicity department of Le Creusot took good care that reports of the victories at Kirkillisse and Lule-Burgas contained references to the inferior qualities of Krupp guns. Things were really serious and for the first time in almost fifty years the head of the firm found himself obliged to speak up in defence of the merits of Essen products.

The sorely tried firm may have found some consolation in the fact that, even if indirectly, they could participate in the business of the victorious side. They were, after all, well in with Skoda, who (the Lord be praised!) were reaping a rich harvest at the very gates of their works, not to mention their interest in Ehrhardt, whose output, to the value of many millions, went through Schneider into Serbia. The coincidence that the German Minister was a certain von Reichenau, while there happened to be a General von Reichenau on the Board of Directors in Düsseldorf, was not altogether mere chance. Finally, Krupp's participation in the Powder Trust made it possible to supply large quantities of German-made explosives to the allied nations opposed to Turkey by a roundabout way—through Paris. "Needless to say," said the reassured *Tägliche Rundschau*, "the German firms approached by the French emissaries refused to supply" whereupon the exports of these self-same firms automatically rose from thirty-eight millions in 1911 to sixty-five millions in 1912.

It was only in the second act of the Balkan drama, after the intervention of Roumania, that normal conditions of the arms industry returned to Essen. They were now openly supplying both sides, for Bucharest was amongst the best customers of Krupp, who had reorganized its arsenal many years before. As Skoda was at the same time supplying both Bulgaria and her opponents, while Schneider and Vickers-Zaharoff were simultaneously engaged in a bitter struggle for the entire Balkan region, the confusion amongst the various interests grew to Babeldom. Each separate case now brought up the question of the actual extent of the European market for armaments, and at this stage commercial machinations connected with the industry unwittingly became sinister political complications.

The most prominent of these is the "Putiloff" affair, beginning with the historic telegram in the *Echo de Paris* January 27th, 1914.

"St. Petersburg: There is a persistent rumour that the Putiloff works in St. Petersburg have just been purchased by Krupp. Should this significant news be substantiated, it will not fail to arouse great misgivings in France. . . ."

Which came to pass. The fact that Putiloff worked under the Schneider patents so that to a certain extent they held the secrets of the French manufacture of arms, sent shivers down the spines of all good Frenchmen. Premier Doumergue cabled the French Ambassador in St. Petersburg to seek official explanations without delay.

Meanwhile enterprising Paris reporters located the right spot at 7 rue de Madrid, the head office of the Schneider company, where they were blandly informed that the firm "have not received any information." But one daring spirit penetrated into the private sanctum of Monsieur Eugène Schneider whom he found to be singularly uncommunicative and embarrassed. He knew nothing beyond what he had "read in the daily papers." But he admitted that if Krupp had really bought the Putiloff works the Germans would undoubtedly obtain access to all the French secrets of manufacture and that the consequences would be extremely serious.

The turbulent echo called forth in the press by the alarming news proved equally futile in finding a cause for the occurrence. Paris opinion was divided between a feeling of indignation at the faithless Russians who had now fallen on the necks of the Germans, and mysterious suggestions that England was at the bottom of the whole business. The London *Times* talked at first of a "Russian trick." A flirtation with Krupp to loosen the purse strings of French investors. Then it became more outspoken and suggested that Schneider himself had raised the issue to secure an opportunity of clearing up the whole complex of Franco-Russian armament connections. Then came statements by Vickers and Putiloff denying such suggestions. And then, to clear up the situation, came a categorical démenti from Essen; they had nothing whatever to do with the whole affair.

What had really occurred? To-day, a quarter of a century later, we have no need to indulge in vague surmise as the entire issue has been cleared up and recorded. The actual events leading up to it form part of the history of Messrs. Schneider, or of Sir Basil Zaharoff, and may be summarized as follows: Putiloff, the largest armament firm in Russia, whose 35,000 employees were always seething with sedition, were in need of money; a trifling matter of sixty millions. This might have been found easily enough had not Schneider's associated bank, the Union Parisienne, already advanced 120 millions to the firm. But Rue de Madrid had other just cause for dissatisfaction. At the last distribution of Russian Government contracts, Schneider came in after both the British and the Germans—obviously due to the machinations of the egregious Sir Basil Zaharoff with his determined striving for influence in the Nicolæff dockyard and the Tsaritsyn Cannon Works. So it amounted to a Franco-British conflict over the Russian market. We've already seen, in the case of Morocco, how divisions of nationality have no relation to big business.

In this case as well Zaharoff-Vickers possessed a French ally in the Société Generale, who were just preparing the flotation of a great Russian loan for the purpose of enabling their Muscovite friends and allies to double the railway lines to the German frontier. Schneider, whose financial resources were, for the moment, heavily strained by the Balkan war, had made efforts to participate in the coming attack on the savings of the Little Man on behalf of Putiloff, but had been turned down.

The doubly aggrieved Schneider's retort was an audacious move; he spread a sensational rumour in St. Petersburg about the acquisition of the Putiloff works by Krupp, which had the effect of a bombshell on the pending loan project of their rivals. The rumour was denied, but the Société Generale fully appreciated the treacherous nature of the ground on which they were moving and the fears created by it were not wholly imaginary. If further misgivings of the potential French lenders were to be allayed, it was essential to make concessions to the dangerous accusers. All parties accordingly made efforts towards a friendly understanding by a suitable contribution to the common cause; the Société Generale's contribution was ten millions towards the financing of Putiloff; Zaharoff's—a basis for an agreement on Russian contracts; St. Petersburg's—large orders to Creusot; and Eugène Schneider's—his silence.

Silence concerning Krupp!

For the worst feature of the false rumours in St. Petersburg was the fact that they were true. Krupp was not merely *ante portas*, but actually right in the great arms factory with its French manufacturing secrets. Ever since the reorganization of the Putiloff works in 1912, nominally carried out under the auspices of Schneider and the Hamburg shipbuilding firm of Blohm & Voss, there was evidence of the financial influence of the Deutsche Bank, Krupp's great ally in the Bosphorus business and in the Far East. The fact that their co-operation extended to Putiloff was established by documents found by Leonidoff in the Russian State archives, in which there are records of the participation of Krupp in the "Syndicate for acquiring the Putiloff works" formed by Schneider and the Hamburg shipbuilding firm, in which the parties concerned "agree to keep one another advised of the tenders for rapidly growing contracts in armament material and to refrain from entering into competition with one another." The contacts became even more intimate towards the end of 1913, when Putiloff required further credits and Skoda (where both Krupp and Schneider were interested) agreed to provide one quarter of the sixty millions required. The balance actually formed the subject of negotiations between Schneider, Krupp, and the Deutsche Bank at the very moment when the pistol shot in the *Echo de Paris* provided Putiloff with another means of raising the needed sum.

Essen interests in the Russian armament undertaking were not affected. Of the twenty-two Putiloff directors, twenty-one were German, as were over half the foremen and charge hands. When the effect on public opinion threatened to create a scandal, the Russian Minister of Marine requested the Managing Director Orbanowski (a German) to become a Russian subject, "in order to prevent further attacks by the Nationalists on the German administration of the works." After a secret conference with the German Ambassador Count de Pourtales, Orbanowski agreed. In other respects everything remained as before and all the key men's posts in the works continued to be filled by representatives from Essen and Le Creusot working together in amicable co-operation. There was, after all, nothing to hide. In fulfilment of the Russian army contracts Putiloff constructed field guns under the Schneider patents and heavy guns under the Krupp ones. The various drawings and metallurgical formulæ passed through the hands of the engineers and experts concerned, irrespective of any consideration of nationality or of national secrets. The particular pride of the French, the "secret" of the Schneider 75 mm. gun, about which all Paris trembled for days, was no exception. Incidentally, the gun was no longer a secret, irrespective of Putiloff. Krupp's agents in any of the smaller States could, at any time, have purchased a demonstration model of one of these guns, although in practice they probably found it simpler to buy one direct from Le Creusot. In point of fact no noteworthy improvements in ordnance metal or in the design of guns, projectiles, or armour, were effected in the decade immediately preceding the Great War which were not passed around all the great armament firms—in consideration, of course, of the customary royalties. André Tardieu, whose business experience in Morocco and the Congo must have taught him a thing or two, referred to this diplomatically in the *Temps*—"Nowadays secrets of manufacture are to a certain extent disordered."

The Putiloff affair remained a muted trumpet call. In other respects the spring and early summer of 1914 presented a picture of weary relaxation, behind which an ominous storm was brewing. The stage was set first in the territory so carefully preserved as a monopoly of Essen since the early days of the Cannon King—in Turkey. It was on the Golden Horn where the big business interests now began to snarl and snap at each other like a pack of wolves.

The apparent cause was a "purely political" matter, the appointment of the German General Liman von Sanders, who headed a mission composed of specially selected German officers sent to reorganize the Turkish army after the Balkan *débâcle*.

Prime mover here was the German Military Attaché to the Porte, Major von Strempel, a gentleman with an extraordinary talent for armament business, who considered it his chief professional task to persuade the Turks to buy obsolete warships and left-over stocks of arms. There was, therefore, nothing to be surprised at when the military reorganization arrangements revealed themselves as merely a skilfully disguised armament ramp of staggering dimensions; the contracts to be placed included the complete rearmament of the Turkish field artillery, the construction of new docks and naval dockyards, heavy guns for the defence of the Straits, and immense fortifications at the Chatalja Line.

The German Mission immediately met with British, French, and Russian opposition of a most determined character. Common action was agreed on in Constantinople and menacing communications were exchanged with Berlin. This war of notes lasted until July, 1914, and it may well be considered the beginning of events ushering in the World War. Contrary to the precedent of Morocco, the attitude of Wilhelmstrasse grew markedly stiffer as the opposition increased. What was it for which it so stubbornly fought on the Bosphorus?

As early as October 28th, 1913, the German Ambassador in Constantinople, von Wangenheim, had submitted a report to the Foreign Office, which gave a clear picture of the crux of the whole matter—

"The local representative of Krupp informs me that the Turkish Government are to conclude an agreement this week with the British consortium proposing to build new docks at Ismid, under which all ships to be built for Turkey within the next thirty years are to be ordered from British shipbuilding firms. At the same time it is reported that France is making her financial assistance conditioned on seventy-five per cent. of all army contracts being placed with French firms."

He recommended "that Turkey be told frankly that we are unable to tolerate any agreements under which the monopoly of such contracts is reserved for other States."

Here we have a delightful picture of what European diplomacy on the eve of the World War described as a "political conflict." They were all in it together again —or perhaps all up against one another: Krupp, Schneider, and the British consortium, concealing Sir Basil Zaharoff, whose card now boasted the names of the combined firm of Messrs. Vickers-Armstrongs, Ltd. The duel of the giants had begun!

The first act was low comedy; one of the combatants sneaked off laden with booty, leaving his ally in the lurch. While the diplomatic representatives of the Entente were still jointly protesting against the proposed military and business plans of the Turks, the British interests concerned had already taken steps to see that their participation in these plans should be definitely assured. The Porte was to grant a monopoly concession for a period of years to a company which they would form for the whole of the Turkish naval contracts and this company was to be entrusted with the maintenance of all the Turkish dockyards and arsenals. It was a master stroke of Zaharoff's, who thereby evened up the score with Paris and St. Petersburg. The receipt was destined to be written only one year later, in the blood of British soldiers mown down by the Zaharoff artillery and machine guns on Gallipoli.

To revert to the dispute over the Military Mission: it went on over questions of army command and similar formalities, whereas it was purely and simply concerned with the sale of cannon. Was it to be understood that the primary issue was political ascendency in Turkey and that any business picked up thereby by national armament firms was to be considered as casual, although valuable, items of secondary importance? The publication of relevant diplomatic records leaves no room for doubt on this point. The French Military Attaché declared: "If France is to participate in the contracts, the German Mission can be accepted." Berlin's reply delivered through the Turks was a curt refusal—meaning "The Krupp contracts are more important for us also than the Mission." This was further confirmed when the overwhelming financial strength of France subsequently enabled her to overcome the opposition of the Porte. The Turkish Minister of Finance Djavid Bey received promise of a loan for 500 millions in Paris and on March 1st, 1914, von Wangenheim warned the German Foreign Office of the probable consequences. "The former position of Krupp is threatened by the possibility of a substantial proportion of the requirements of the Turkish army, in the shape of its entire field artillery rearmament, being lost to Schneider-Creusot." On this point, however, Berlin remained adamant; towards the end of March it was decided that, in the event of any injury being done to Krupp's interests, the entire Military Mission, for the sake of which the Government were prepared to run the risk of war, should be recalled!

Although an actual break was avoided, the situation grew appreciably worse. Another bone of contention was presented by the Bagdad railway, the vast undertaking of the Deutsche Bank, the bulk of the permanent way material being supplied from Essen. When Djavid Bey, under pressure from Paris, declined to provide security for the Bagdad-Basra section, the Deutsche Bank delayed its quit claim to the last section, Basra-Koweit, which affected British interests unfavourably. On July 18th von Wangenheim reported to Berlin that the German contractor "threatens to cease work on the railway, while Djavid retorts by threatening to transfer the Krupp contracts to France." This naturally caused the Ambassador to issue an immediate protest. "I informed the Grand Vizier that it is his duty to intervene immediately in order to protect higher State interests from injury through Djavid's stubbornness."

"Higher State interests" in July, 1914; securities for loans and contracts for cannon!

The dispute was acquiring the character of a Turko-German conflict, but the chief cause should not be overlooked: it was the rivalry between Schneider and Krupp, which after temporarily slowing down, had suddenly flared up with dangerous violence. And this in a situation where Russia was watching the whole trend of German policy in Turkey with deadly irritation, while Great Britain was feeling far from comfortable over the pending completion of the Bagdad railway. When the fatal shots were fired in Sarajevo, the negotiations in Constantinople had practically broken down. The quarrel over the "Krupp Cannon Mission" led directly to the World War.

ON THE EVE

What was the state of things in the "Armoury of the Empire" on the eve of that hour which was to give meaning to the effort of a half-century? What did the name of Krupp conceal, its dull ring threateningly suggestive of the voices of its giant cannon?

The Krupp Works were Germany's greatest industrial enterprise, with ramifications embracing every aspect of steel production and application, but compactly homogeneous with its thousand iron ore mines, collieries, ironworks, and factories on the Rhine, in Essen, Annen, and Magdeburg, and shipyard at Kiel.

The range of products included every item directly or indirectly connected with the "bloody trade"; guns, projectiles, fuses, rifle barrels, armour plate, gun shields, wheel tires, ships' fittings, engines, rolling plant, and steel ingots.

The development of the manufacturing resources had

been carried out on a gigantic scale in every respect and now included—

18 rolling plants	81,000 employees
53 Martin furnaces	2½ million tons of coal used annually
180 steam hammers	
430 steam boilers	3½ million tons of ore used annually
550 steam engines	
1000 cranes	63 million tons of water used annually
7200 work machines	

The works used more gas than the city of Essen and as much electric power as the whole of Berlin. The total length of the railway lines inside the works would have sufficed to connect Frankfurt and Munich, while the telephone lines would have reached from Strassburg to Königsberg.

The firm maintained four artillery testing grounds larger and better equipped than those belonging to any government, and the test firings of their guns and rifles were like a private war; the 62,400 rounds fired in the course of the preceding year required four freight trains of thirty fully laden trucks each to transport the 2500 tons of projectiles and explosives.

The works had turned out the biggest cannon in the world, the huge 17-inch gun nicknamed "Big Bertha" in charming compliment to the owner of the business; its projectile was fired with a muzzle energy equivalent to that of five express trains each weighing 250 tons and travelling at a speed of over sixty-two miles per hour.

The total sales of guns up to the end of the previous year amounted to 24,000 to the German Government, and 26,000 to fifty-two foreign countries, twenty-three of these being in Europe, eighteen in America, six in Asia, and five in Africa.

In a word: Krupp was the pulse of Germany's armament industry, beating in anticipation of the "great hour."

On the eve of such events it was more important than ever for the Essen firm to extend their activities beyond the mere production and sale of arms. Although the Service Estimates had swollen to two and one-half billions, the principal beneficiaries were faced with a growing problem. German public opinion had not yet been subjected to a sufficiently intense degree of military hypnotism, to preclude any possibility of an outcry against the excesses of the armament craze. For this purpose the "Defence Association" was founded, an all-German organization having for its object an intensive propaganda campaign for the creation of a huge army and for popularizing the idea of a war of aggression. The head of this organization was General Keim, a tried and true Krupp agent. And here again the firm endeavoured to conceal their financial support of the association. That the latter's press organ, the newspaper *Überall*, had sixteen whole pages of a single issue taken by advertisements of Krupp and of the Schichau shipyard, was a little mistake which was promptly rectified.

The establishment in 1909 of the political secret service fund by the Union of Industrialists, at the instance of their President, the then managing director of Krupp, Rötger, marked the beginning of the movement to influence the long ignored Reichstag. From that time on only odd rumours of the Nibelungen Hoard of the steel magnates were heard, but the effects of its financial power were seen clearly enough. The constituencies of unruly members of the Left Wing became the scene of increasing inspired demonstrations of patriotic electors to whom in the majority of cases the alarmed mandatories hastened to capitulate. The object of these manœuvres is betrayed by a proclamation issued from Essen during the elections of 1912 and signed by Stinnes, Kirdorf, and the Krupp Director Gillhausen; it demanded nothing less than the abolition of the secret ballot, to reduce the strength of the Social Democrats by a third. Pressure was also brought to bear on the universities, through a pseudo-scientific association under the leadership of the Krupp Director Klüpfel. Its purpose was to finance so-called "Tendency Chairs" to off-set the liberal views prevailing in most of the faculties by bringing in professors of more nationalistic mentality who would be charged with the duty of inspiring their colleagues with the patriotic ardour so important for the well-being of big business.

No cannon without the press! The firm acquired a controlling interest in the official Wolff Telegraphic Bureau, that it might serve as the mouthpiece of Essen where and when required. The Government gave its tacit consent to the arrangement. It had become necessary to withdraw from the *Berliner Neueste Nachrichten* as the connection between its political aims and the business interests of its proprietors had become too obvious, but the management of the paper remained as before, somewhat less suspiciously, under the auspices of the Union of German Iron and Steel Manufacturers. An interest was also acquired in one of the leading dailies of the capital, the *Berliner Lokalanzeiger*, whose proprietor, Scherl, was in financial difficulties and glad of help from Krupp. It was this same *Lokalanzeiger* which, during the critical hours of August 1st, 1914, was to publish the "premature" report of the German mobilization, thereby precipitating the war avalanche. Krupp's interests in the press were not confined to Germany. Among the foreign publications in which they held an interest were the *Etoile Belge* and the *International Review of All the Armies and Navies.* There were hundreds, if not thousands, of other editorial offices in which the influence of Essen was more or less decisive; where a "bonus" for small journalists did not suffice, a large-scale purchase of advertising space from the publishers was invariably effective.

The headquarters of this network was the Intelligence Office in the Essen Administration building, termed the "Public Opinion Control Office" by a well-informed critic. This organization, half bribery centre and half espionage bureau, was entrusted with the important task of collecting information relating to armaments, ordnance matters, new inventions, and similar subjects appearing in any German periodical or in any of the leading foreign papers. This information was then supplemented by special methods. The object was twofold—the collection and distribution of intelligence and the penetration of enemy secrets while concealing one's own through the dissemination of false or irrelevant information relating to them. Although the work was carried on anonymously and with the utmost caution, the smoke screen was occasionally lifted; in the spring of 1914, Liebknecht reported to the Reichstag: "In 1910 the Krupp Intelligence Office secured possession of the latest barely completed drawings of new ships for the Austro-Hungarian navy and it also figured in a bribery scandal in the Argentine and in an unpleasant affair in Sweden." We already know the department's achievements in

Brazil, Chile, Turkey, China, and, last but not least—its brilliant performances in Berlin. It is, therefore, not surprising to learn that the expenditure of this versatile department exceeded a million marks. The audited accounts of the firm for 1913 comprised eighty lines of print, but any reference to this item would be vainly sought for in them.

Public opinion, Reichstag, educational establishments, and press—even all these did not suffice. Krupp's, with 80,000 workmen and employees on July 1st, 1914, found it essential to establish direct points of contact in the increasingly dangerous labour union field of battle. However unimportant the Socialist Left might be politically, they nevertheless succeeded in waging an energetic guerilla campaign against the day-to-day practices of Essen, in the course of which many an awkward matter was brought to light. A statement, made for publicity purposes, that the firm's pay-sheets showed increases of wages amounting to seventeen per cent. within five years, was countered by proofs that the cost of living during the period in question had gone up nearly thirty per cent. The fact that the incidence of sickness amongst Krupp employees averaged sixty-seven cases per hundred men, coupled with 5000 accidents in the works in a single year, indicated that conditions prevailing in the Krupp establishments were not what they should have been. There is also evidence that the patriarchal tyranny dating from the days of the Cannon King was still exercised in more or less the same way; whereas thousands of other German industrial establishments already had fixed scales for wages, the Essen administration consistently refused to enter into any discussion of the matter and ignored all requests for negotiations. But they were all the more eager to suppress any criticism on this point and displayed extraordinary lack of foresight in supposing that the demands for social justice made by their employees could be put off by such a crude trick as the one to which they resorted.

This trick was termed a "factory union"—a pseudo trade union organized by agents of the firm and kept alive by standing subsidies. In other words, the American "Company Union." Its avowed object was the maintenance of harmonious relations between employees and employers, but its pretended independence was an empty farce; the newspaper *Nationaldemokrat* proved that "nothing is done without the approval of the Board's Secretary Halbach, who takes all his instructions from the Director Vielhaber." To develop this fictitious movement throughout the whole Empire, Krupp financed a central office in Berlin, with publicity experts drawing up to 6000 marks in salaries. A rival organization of trade union idealists, who genuinely believed in the disinterestedness of the movement, was promptly broken up by Krupp agents. The autocrats of Essen insisted on blind and unconditional obedience from those whom they paid to an extent which was not fully attained in practice until twenty years later.

But let us return to the works, the hotbed of all the subtle ramifications of the business here disclosed. It is not easy to draw a clear picture, for the subject does not stand still. On the eve of the War, Krupp was again in the midst of a programme of feverish expansion. To use a traditional metaphor of the Essen firm, the dramatic was changing into the idyllic. In 1911 precisely the same tactics as were employed to overthrow Gruson in 1892 were employed for the capture of the Westphalian Wire

Works in Hamm. In the words of Krupp von Bohlen it was desired "to ensure continuity of employment for our steelworkers for years ahead, without having recourse to the establishment of new factories in competition with existing ones." The real cause was the fight against Thyssen and the Phœnix concern over the quotas in the Steel Manufacturers' Association. As the statutes of the syndicate precluded direct horizontal expansion, the great producers endeavoured to get a footing in industries consuming their products in order to increase the scope for their output. On this occasion Krupp's victims were the works in Hamm, Germany's leading makers of wire, which were captured by the time-honoured method of buying up blocks of shares through dummies until a controlling interest had been acquired and the business was pitilessly swallowed up.

The final act of this development occurred late in the summer of 1914, when Krupp concluded agreements with the Thommé-Werdohl Wire Works and the wire merchants Messrs. Kunne of Düsseldorf, through which both these firms came under Krupp control. The Essen firm had thereby established itself in a leading position in a further field of steel consumption. The coming boom in barbed wire, with closed frontiers and lines of trenches from the Channel to the Alps, from the Baltic to the Black Sea, was to find Krupp well equipped to deal with it.

The expansion of this gigantic concern, resumed after ten years of internal reorganization and consolidation, did not stop here. Following the example of Thyssen, somewhat hesitatingly at the outset but with increasing energy after 1913, Krupp pushed forward into the French iron-ore districts. Activities in this direction were camouflaged by the employment of a Dutch firm of wholesalers, Messrs. de Porter, intermediaries for securing mining concessions, two in the Department of Calvados and two in La Manche. They also purchased a large tract of land in Lorraine between Maizières and Woippy, where it was proposed to erect works. About this same time Krupp secured exploitation rights from the important graphite works of Kallowitz in Bohemia and acquired sixty more collieries on the Westphalian property of the Duke of Croy.

This last huge development was carried on by the boom in armament industry in 1913-14 which put all earlier ones completely in the shade. "Krupp can barely cope with all the orders pouring in" was the Essen report in the commercial press. The dividends of the Krupp company scarcely reflected the fantastic profits extorted from their customers as, on the admission of the Essen firm's responsible representative Dr. Schmidt, "considerations of the rapacity of the employees and the growing hostility displayed towards the firm in industrial and Government circles" made it essential to keep dividends as low as possible. But the writing down of assets, appropriations to reserve funds, and secret adjustments of accounts could not wholly conceal the profits made, as can be seen by dividends declared during the three years preceding the War—

1911—10% equivalent to 18 million marks.
1912—12% equivalent to 21 million marks.
1913—14% equivalent to 25 million marks.

Essen achieved a record here, as the twenty-five millions paid in dividends in 1913 were the highest sum ever paid to the shareholders of a German limited company. While on this subject it may be interesting to quote an extract from "The Millionaires' Annual" for 1913, showing the fortunes of the five wealthiest persons in Germany in the year before outbreak of the War:—

	Marks
Frau Bertha Krupp von Bohlen - -	283 million
Prince Henckel von Donnersmark -	254 "
Baron von Goldschmidt-Rothschild -	163 "
Duke of Ujest - - - - - - -	154 "
Emperor William II - - - - - -	140 "

The greatest fortune, the highest dividends, the largest business concern, the most formidable gun—all this was Krupp's on the outbreak of the World War. If their chief was to be believed, also the clearest conscience!

THE WAR

Did Krupp anticipate the outbreak of the War? One important witness states that they did: Dr. Wilhelm Mühlon, a director of the Krupp A.-G. from 1911 to 1915 and therefore in a position to know. On the recommendation of the Imperial Chancellor Bülow, this young barrister was first employed by Krupp von Bohlen as his private secretary and was subsequently nominated to the Board of the company in 1911. Mühlon rendered valuable service in the Morocco trouble, and during the early part of the War he did some good diplomatic work in Roumania. His activities on behalf of Krupp abruptly ceased at this point and he was next heard of in Switzerland, where the critical atmosphere of a neutral country changed the former Krupp director into a bitter opponent of the War.

Mühlon made alarming disclosures to Mr. Herron, the confidant of President Wilson, and to the press. He stated that six months prior to August, 1914, Krupp's received secret advice about the coming war from Berlin and thereupon proceeded to reconstruct the works accordingly.

As long as free discussion was permitted in Germany, *i.e.*, from 1919 to 1932, the firm did not find it necessary to repudiate this accusation. But when conditions for safeguarding secrets of the armament industry became more favourable in December, 1933, Krupp's broke their silence. In the course of the trial of the Catholic leader, Professor Dessauer—accused, amongst other things, of friendship with Mühlon—Krupp von Bohlen made a sworn statement, which is reproduced in the form in which it appeared in the *Rheinisch-Westfälische Zeitung:*

The Public Prosecutor then refers to the lying statement of Muhlon, that the firm of Krupp knew war was imminent six months before it began.

Mr. Krupp declares under oath that his firm did not receive any such information from the Government. He explains that it is the practice in all countries to make provision for war eventualities when placing contracts with armament firms. His company received no advice beyond this, before the outbreak of the War.

Public Prosecutor: "Did we go to war without special preparations?"

Mr. Krupp: "I am definitely under the impression that

such was the case. . . . The shortage of explosives in 1914 resulted in great loss of life for us at the front."

Notwithstanding its formal and definite character, this explanation is inadequate. If Krupp's did not receive such information, is it not more than likely they received due warning of the imminence of war in some other way? Were there not revised and accelerated dates stipulated for the deliveries of war stores, which led to the extensions of the works referred to by Mühlon, in the spring of 1914? This suspicion is confirmed by the vagueness of the reply given the Public Prosecutor's question about "special preparations." Although he was well aware of the exact nature of the accusations levelled against his firm, Herr von Bohlen contented himself with making a wholly irrelevant reference to the shortage of high explosives experienced by the German army in the autumn of 1914. Not a word about the preparations made in his own works!

He had good reason to say nothing, as Mühlon's statement is supported by evidence which cannot be refuted by any retrospective sworn statements—the records of imports into Essen of certain raw materials, essential for the manufacture of gun steel. The most important of these, nickel, comes from New Caledonia, and Germany imported about 3000 tons per annum of it until 1909. After this, the imports went up rapidly and totalled about 20,000 tons for the four and a half years immediately preceding the War—the whole of this nickel going to Essen. The same thing applied to ferro-silicon. Krupp concluded an agreement with the International Ferro-Silicon Syndicate controlled by the Frenchman Giraud-Jordan, in which Article 10 laid down that only a war between at least three of the European Powers was to be considered *force majeure*, and that a war "merely" between the countries of the contracting parties was not to be considered as such. From 1912 onwards Krupp required a further 1000 tons per annum of this important material and secured prompt delivery of it, while French business associates wrote to the firm significantly referring to "due regard to the quantum specified in the event of war." In the case of cyanamid and Spanish and Swedish ores, the quantities discharged at Rotterdam in the year before the War showed an alarming increase. The firm were obviously in a hurry to lay in huge stocks of war material, a procedure which would be far too costly to indulge in unless inspired by definite information and governed by fixed dates. Statements made by Krupp engineers indicate that in the spring of 1914 not only were the stores and warehouses of the firm filled to their utmost capacity by the addition of "emergency supplies," but that the machine tool equipment of the shops and their organization, etc., were carefully calculated and planned to deal with the possibility of a sudden demand for a greatly increased output.

Then again, why the decision made in the summer of 1914 to raise the share capital of the company from 180 to 250 millions? Since 1913 there had been signs of a universal depression following on the boom, so why should Krupp think it desirable to take in seventy million marks of additional capital? The adoption of the German Navy Acts at the beginning of the century and the commencement of Dreadnought construction in 1908 were, in each case, preceded by large-scale extensions of the Krupp works. The connection between the Berlin reference to imminent danger of war and the financial dispositions of the Krupp directors was too significant to be regarded as a fortuitous coincidence.

Were the Essen directors forewarned of the War? Undoubtedly the firm did not hesitate to repudiate the documentary proofs of this, any more than German statesmen of 1914 hesitated to do so, but Krupp's did not, in any case, venture to deny that they were fully prepared for war.

The thoroughness and extent of their preparations may be indicated by the clockwork precision of their activity through the critical days of August, 1914, during the period of mobilization, with millions of men called to the colours, the declaration of a state of siege, and the closing of the national frontiers, thereby severing all contact with the hostile countries overnight. None of these things affected the business of the armament firms. The deal between Krupp and the New Caledonian firm "Le Nickel" was carried through with exceptional smartness. The Board of Directors included Baron de Rothschild (of Paris), Sir Basil Zaharoff, and two Germans acting for Krupp, representing 210 shares in the company. It is known that the agreements concluded by this predominantly French company with Essen foresaw the possibility of war and provided for the diversion of shipments which might be affected thereby through neutral countries—preferably via Norway. This was done, as may be seen from the statements made by Senator Gaudin de Villain in January, 1917, and Deputy Henri Bérenger, speaking in the Paris Chamber on January 24th, 1919—

"In September, 1914, that is, after the outbreak of war, the Norwegian full-rigged ship 'Benesloet' took in a cargo of 2500 tons of nickel in Freisund, New Caledonia, which is French territory. Half the shipment was paid for in advance by Krupp, the consignee. While at sea the Norwegian vessel was stopped by the French armoured cruiser 'Dupetit-Thouars' and brought into Brest early in October. A prize court declared the cargo of nickel to be contraband of war, but an urgent message from Paris ordered the immediate release of the ship. The astonished local authorities and the prize court declined to act on the order, until the Ministry, at the instance of the 'Le Nickel' Company, formally confirmed the order of release. On October 10th, 1914, the 'Benesloet' resumed her voyage and . . . her cargo duly reached Essen via Norway."

The indignant French patriots who levelled this accusation and other similar ones against traitorous contractors in the Chamber must not imagine that such practices were peculiar to their country. In the summer of 1915, the following report came from England:—

"London, June 20th (T.U.). The metal merchants Hetherington and Wilson of Edinburgh, who made deliveries of iron ore via Rotterdam to the firm of Krupp after the outbreak of war, were each sentenced to six months' imprisonment and to a fine of two thousand pounds."

It was, however, exceptional for cases of this kind to lead to a conviction in the courts, and other British exporters of hundreds of tons of copper and nickel, sent to the Central Powers via Sweden, went their way unexposed.

Precisely the same thing applied to Germany, where on February 3rd, 1915, the official Wolff Telegraph Agency reported that considerable numbers of shell-turning lathes of German manufacture had been sold to

Great Britain and to Russia, via Sweden. No steps were, however, taken in the matter and until the late autumn of 1916 the export of ordnance steel was permitted notwithstanding the protests of the Ministry of War. Switzerland alone took up to 250,000 tons a month in exchange for other goods, the risk of loss due to exchange fluctuations being covered by the Reichsbank. The chief culprits in this shameless trade plied behind the backs of the bleeding nations were Thyssen, Stinnes, and Krupp.

To come back to August, 1914, and the Essen works in the greatest hour of their history. It is interesting to note the functioning of this machinery which provided the main driving force for Germany's onslaught against half the world. Despite the rigid military control which imposed secrecy on everybody and made all business and technical detail confidential matters of State, enough can be seen to realize that an immense concentration of all available energy had been made.

On receipt of the secret instructions from Berlin full speed was ordered in all the hundred odd departments of the immense works. A feverish activity commenced in all the great workshops, while, in accordance with careful preparations long since made, anti-aircraft detachments mounted their searchlights and A.A. guns on the tower of the Head Office and all around the works. Surrounded by a cordon of military pickets and detectives, the roar of 10,000 machines continued day and night, week-days and Sundays, turning guns and projectiles, rolling armour plate, shaping torpedo bodies, and building gun mountings. Krupp's also adopted the slogan "The greater the foe the greater the honour," and they were determined to prevail in the contest against Sheffield, Le Creusot, Petrograd, and Pittsburgh combined. The contracts placed by the Government were unlimited and Berlin would take whatever Essen could deliver, which in the first year of the War proved to be about two and a half times as much as the output of the preceding twelve months' boom. Even that did not suffice and the cry was always for greater and quicker production. In barely twelve months thirty-five large buildings were erected and filled with machinery to increase output. The number of men employed rose from 79,000 to 95,000. Most of them were exempted from mobilization, as it was considered to be just as important to manufacture guns as to fire them. Such gaps as did occur were immediately filled with hurriedly trained women, whose numbers swelled to 5000 within a few weeks. The energy of the industrial giant seemed inexhaustible, in the midst of turmoil of the first year of the War, the rolling plant of Messrs. Capito & Klein of Dusseldorf was acquired and incorporated in the Krupp concern.

A truly great epoch had dawned. Heroes fell and profits rose. Financial results of the first year of the War exceeded the most optimistic anticipation—they showed a gross profit of 128 million marks, and although the accounts were cooked to minimize the profits made, this figure was so imposing that it required some skilful juggling by the auditors to reduce it further. This will be referred to later. The auditors contrived to show that a dividend of "only" 12 per cent. was available for distribution, which was no more than in the last year of peace. As the firm were shrewd enough to increase their share capital to 250 millions, the total sum represented by the dividends was thirty millions after ninety-eight millions had been accounted for by appropriations to reserve and writing down of asset values. This is the Krupp family's modest record for the first year of the War—recognized by the award of the Iron Cross, First Class, which H.I.M. the Emperor, in recognition of the war services rendered by the firm of Krupp, was graciously pleased to bestow on Herr Krupp von Bohlen on the occasion of his visit to Grand Headquarters and to invest him with in person.

It is a great thing to be termed a patriotic firm. This distinction was well earned by the production of the 17-inch gun which knocked the Belgian and French forts to bits. The honours bestowed on Mr. Krupp and the creator of the monster gun, Professor Rausenberger, by Bonn University in the form of honorary degrees as Doctors of Philosophy, were gratefully accepted. Philosophy? There was certainly need of some, if one was to retain a clear head for business amidst all these storms of patriotism, particularly for that class of business which did not bear shouting about.

Meanwhile Krupp had regularized their position in regard to the International Powder Trust, in which they had substantial holdings and were represented on the Board of Directors by Captain Dreger. An exchange of shares had been effected in order to sever connection with their British friends without infringing the law or incurring any risk of sequestration of enemy property for either party. The secret negotiations in connection with this matter were conducted in a neutral country and scrupulously avoided touching on any controversial points arising out of the War.

During all this time heavily laden freight trains, carrying vast quantities of high-grade steel of Krupp make, continued running to Switzerland. Its ultimate destination could easily be guessed; it required little imagination to do so, any more than it did to discover the exact purpose of the "goods stations" or little depots, in which the Krupp name and trade mark were filed or ground off the steel bars before these were sent on to France. If any awkward questions of high treason or of trading with the enemy were to arise, a defence could always be made before the Supreme Court in Leipzig that these shipments were counter-balanced by imports of raw materials essential for war purposes, which was true enough. Nobody in Essen or in Le Creusot had the slightest intention of assisting the enemy; it was merely a question of ensuring that the War should not be ended prematurely through a lack of materials, and with this object in mind there was a tacit understanding that both sides should play into one another's hands.

Should rumours of these transactions have become too insistent, Krupp were to maintain a discreet silence. This applied in particular to the little matter of the shell fuse patents which appeared to defy all attempts to keep it secret. The German soldiers in the trenches of Northern France might, perhaps, have failed to appreciate the meaning of the cabalistic mark KPz 96/04 on unexploded British shells which they had picked up. But it happened to be the technical designation of the Krupp patent fuse ("Krupp-Patentzünder") for which licences were acquired by Vickers and other British armament firms before the War and which only very few people were likely to have identified as such if they had happened to read the following report in the issue of the Berlin newspaper *Vorwärts* dated April 27th, 1915:—

"London—In the House of Commons Lord Charles Beresford asked the Prime Minister if it were true that the Government was paying a royalty of one shilling per fuse on a German patent to the firm of Krupp; if so, whether this money is paid by the British armament firms to the Custodian of Enemy Property for the Krupp account and whether the Prime Minister would state the amount of the royalties payable to Krupp in respect of the ammunition expended at Neuve Chapelle. This question is down on the Order Paper of the House for Wednesday."

No further reference was, however, made to the matter in the German press, being suppressed by the censorship—apparently at the instance of Krupp. The reply quoted in *The Times* of May 5th, 1915, indicated that the British Government were inclined to display caution in dealing with the question and were unwilling to commit themselves—

"House of Commons proceedings of May 4th—Mr. Barker: 'The Government's agreement relating to the payment of royalties of 1s. 3d. per fuse manufactured in this country under the British Krupp patent, expired on July 16th, 1914. Since that date no royalties on any of the fuses used have been paid."

Of course, nothing was paid. But the gentlemen keeping the accounts of Vickers-Armstrongs, Ltd., conscientiously booked "K" on the debit side of their books, and every month they credited it with 1s. 3d. for every shell which had left their works.

At the end of the first year's fighting and the failure of either side to achieve a victory despite the enormous sacrifices made, protests began to make themselves heard in the ranks of the suffering people. The beneficiaries of these wonderful times were alarmed at the thought of a sudden conclusion of peace and in their anxiety to avert such a possibility, the worthy armament makers on both sides were in entire agreement. Sir Basil Zaharoff expressed the hope that the War would be fought to the bitter end and precisely the same demand was made by Krupp. Their Managing Director Hugenberg and former Chairman Rötger, who represented the firm and held prominent positions in the leading industrial associations of Germany, added their voices to that of Stinnes in clamouring for a "victorious peace."

Nor did they hesitate to call things by their right names. In a memorandum addressed to Imperial Chancellor von Bethmann-Hollweg in December, 1915, by the West German Steel Manufacturers (*i.e.*, war contractors) demanding "persistence to a victorious ending," the following unmistakable passage occurs: "The resources available for the industry of Germany are such as to enable it to supply all the munitions and war stores that may be required by our valiant troops and by those of our faithful allies for many years to come."

They wished to go on supplying those things "for many years to come," the worthy Krupp, Stinnes, and Thyssen, and they had taken care to raise the price of steel and iron so as to impose an additional burden of 100 million marks per annum for shells alone on the Reich. The idea of what was to constitute the "victorious ending" was defined in the famous "Memorandum from the Six Associations" in which the Essen firm played a conspicuous part. It was quite a modest idea: In the East, Poland, Lithuania, and the Baltic Provinces were to be "liberated," *i.e.*, annexed by Germany, and a slight adjustment of the western frontiers of the Empire was to be made to include the north-east corner of France from Belfort, Verdun, and the mouth of the Somme up the Channel, an adjustment which would dispose of the question of Belgium. Needless to say, the inhabitants of the conquered territories would, after the War, "have no influence in regard to the political destiny of the Empire"—a status which can only be likened to that of a colonial dependency.

The whole question as to what was to happen after the War was a bugbear to the gentlemen of the steel and coal industries. Were the millions of people then shedding their blood in silence at the front going to demand political rights for doing so? In Essen, where an extraordinarily keen perception of danger signs had been developed, they seemed to discern the bogy of a movement towards democracy. Already in August, 1915, soon after the Emperor's proclamation of a general political amnesty and his appeal for a party truce, a warning was addressed to the Government and the Supreme Command by the Managing Director of Krupp and the Secretary of the Colliery Association von Löwenstein, in the following terms:—

"If after this war German industry should be exposed to the perils of extreme trade unionism on the one hand or to the increasing tendency for State interference with business interests on the other . . . we shall have to face the risk of seeing our industrial well-being collapse as rapidly as it developed."

This method of dealing with an awkward political development by threatening bankruptcy was quite in keeping with the time-honoured Krupp tradition of the days of the Cannon King, but the firm did not content itself with mere petitions to the authorities. Backed by the untold millions of their war profits, which they found it increasingly hard to conceal, they began to create a system of press and public propaganda aimed at diverting the coming struggle against "the internal enemy." Although this organization subsequently became independent and identified with the name of Hugenberg, this did not affect the historic fact that it was founded in Essen and was financed by Krupp. It originated from a conversation held in the winter of 1915 between the Prussian Minister von Schorlemer and a small group of Rheinish-Westphalian industrialists, which included Krupp von Bohlen and Emil Kirdorf. When the latter were offered an opportunity of acquiring the Scherl publishing concern, the most important press undertaking in Berlin, Krupp jumped at the offer and appointed his Managing Director in charge of the new publicity organization which then began to extend the Scherl concern's business in all directions. The Ala Advertising Agency was used to exercise pressure on the irresolute Left Liberal provincial newspapers, and the official Wolff Bureau was faced with competition from the specially created Telegraph Union as a still more effective means of spreading unscrupulous reactionary propaganda among the public. The still young film industry was penetrated by the establishment of the "Deutsche Lichtspielgesellschaft" (German Motion Picture Company), the forerunner of the Ufa Company. During the years of the War the great industrialists succeeded in building up an immensely powerful propaganda machine, the effect of which was soon felt by the unsuspecting parties of the Left. The principal financial backers were, of course, the House of Krupp, and it was the Essen millions which rendered possible the beginning of a movement which, after seventeen years of intensive reactionary propaganda, led directly to the event of January 30th, 1933.

The year 1916, which saw Germany's furious battles on both the Eastern and Western Fronts, the entry of Roumania into the War, the taking over of the Supreme Command by Hindenburg and Ludendorff, was a year of high hopes for Krupp. In the works, still being extended on a phenomenal scale, there were 118,000 employees including 20,000 women and girls, who were both useful and economical as they received only half the pay of the men although their sensitive touch made them particularly useful on certain classes of work—such as the production of shell fuses. Krupp's activities exceeded the most fantastic speculations of the most imaginative experts of the General Staff. All the data relating to pre-War production was now discarded and the output of guns was twelve to fourteen times, that of projectiles twenty times as great as, the output provided for in the mobilization plans. One single shell-turning shop dispatched 20,000 shells to the front day after day.

Maximum output? The word is purely relative in the Krupp Head Office and the "many years" until a "victorious ending" demanded a continual rise in production. A large tract of ground had just been purchased at Freimann, north of Munich, where it was proposed to erect a new gun factory to be completed by 1917. It would serve to absorb a further twenty-five millions of war profits and would enable the General Staff's strong recommendation for the decentralizing of manufacture to be carried into effect. In the newly established "Bavarian Gun Works, Fried. Krupp" Essen was, for the first time in seventy years, associated with Arthur Krupp-Berndorf. The Deutsche Bank was also interested in the undertaking which went further than the Munich establishment, Berndorf enabled contact to be made with Austria, where a secret competition of expansion-hungry German firms had already begun.

The west offered a still greater attraction in the shape of the industry of Belgium, together with all the resources of that unfortunate country. Its collieries, ironworks, railways, gas, water, and electricity had all been taken over by the Army of Occupation. For a long time the Ruhr industrialists headed by Krupp and Stinnes had been casting envious eyes on this appetizing morsel, but the Berlin authorities declined to accede to their wishes in order to defer any action which would rightly be regarded as the first step towards annexation. In 1916 the hesitating Bethmann-Hollweg was ready to give way and approved a scheme for the exploitation of Belgium's industrial resources "as a retaliatory measure against the economic blockade," but which was equivalent to a forcible seizure by the Ruhr magnates. While the political situation of Germany was thereby rendered substantially more difficult, the interested parties cautiously proceeded with their scheme, hiding behind organization in 1916 of three limited companies in Essen, the Transport Company, the Industrial Company, and the Land Company. These companies were granted a virtual monopoly for the exploitation of the means of transport, industrial resources, and land values of Belgium. The principal shareholders were Krupp, Stinnes, Thyssen, and Haniel, who all took good care to keep unwanted outsiders at arm's length.

The owners of all this property (most of whom were fugitives) did not, of course, get a cent. Their factories, etc., were assessed at an official value, barely half the real one. That official assessors then proceeded to enter the service of the fortunate recipients of these gifts aroused no surprise. Even the modest amounts corresponding to the valuations were not paid by the "purchasers" in Essen, Mülheim, and Duisburg in cash, but by bills of exchange endorsed by the Prussian State Bank and payable six months after the conclusion of peace. Meanwhile the earnings of the properties amounted to millions. The new owners were not concerned with any thoughts of running the various undertakings on a sound basis which might lead to competition with their own interests. Their idea was merely a systematic robbery of all they could get their hands on. In most cases this involved the barefaced plundering of the buildings of everything that could be stripped from their bare walls. Complete installations of machines, tools, etc., were dismantled and shipped to Germany, where the shortage of materials of all kinds enabled them to be sold at fancy prices. The burden of the economic blockade was borne by the German nation while the result of the "retaliatory measures" could be credited to the accounts of Krupp, Stinnes, and their friends.

These transactions enjoyed the special favour of the military censorship and not a line about them was allowed to catch the eye of people who might possibly begin to suspect that demands for the annexation of Belgium, put forward by the great industrialists, might be inspired by motives not wholly connected with ardent patriotism. Furthermore, Krupp's faithful friend the Emperor was always at pains to remind people of the great services rendered to the nation by the Essen firm. After the Battle of Jutland he sent a telegram to Hügel, as follows:—

"Wilhelmshaven, June 5th, 1916—Herr Krupp von Bohlen und Halbach, Essen.—As an immediate result of the impression made on me by eye-witness accounts of the battle in the North Sea, I wish to place on record that our success was due to our excellent guns and armour and more especially to the destructive effect of our shells. The battle is, therefore, also a day of triumph for the Krupp works.—William R.I."

It is true enough that the guns and armour of the German ships were made in Essen, but it is equally true that the whole of the armour of the British boats was manufactured under the patent of the Krupp Director Ehrensberger and that the British shells were fitted with Krupp fuses. Neither must it be forgotten that the optical instruments on board the British ships were made by the firm of Gœrtz-Anschutz associated with Krupp, ostensibly for a Dutch Maatschappy who duly shipped them over to Britain. William II's congratulations were well deserved, even if the controversy amongst naval experts about the Battle of Jutland makes it a matter of opinion as to which of the combatant sides found it "a day of triumph" for Krupp products.

At the end of the second year of the War the Essen firm was again faced with the ticklish problem of divulging the brilliant results of the year's work to a suffering and starving nation. On this occasion the gross profits totalled 143 millions, which had to be made to disappear if the popular clamour for the drastic taxation of war profits was to be scotched. The profits were, therefore, reduced to sixty-three millions and a further fifty millions were put in reserve that the same dividend of twelve per cent. might be announced as in the preceding year.

The Krupp family therefore received another thirty millions on their 250 million share capital.

If the working of war industries could proceed so brilliantly and smoothly, the war itself might go on. The smoothness of their working was certainly surprising. War histories, crammed with accounts of deeds of valour, have little to relate about operations directed against industrial establishments belonging to the enemy. The aviators who bombed schools and processions of children behind the enemy lines appeared to make a point of avoiding attacks on certain industrial centres. So far as the establishments at Essen and Le Creusot were concerned, their distance from the Front and the relatively limited radius of action of contemporary aircraft may account for their immunity from air attack. But just behind the lines of Allied and German trenches in Northern France were important industrial centres which were of particular value to the respective combatants as sources of supply for raw materials. These centres were the iron-ore mines of Briey on the German side and the Pas-de-Calais coalfields on that of the Allies.

As regards the first named, occupied by Germany at the beginning of the War, their output of millions of tons of ore per annum, together with that of the adjacent Lorraine territory, supplied ninety per cent. of the requirements of the German arms factories producing gun steel. It has been stated on the highest economic authority, that if these sources of supply for raw materials had not been available for Germany, she would have been hard put to maintain her production and the War would probably have ended in 1917.

It is astounding to note that although the Briey mines and ironworks were less than twenty miles behind the

German lines and therefore well within the range of French bombing aircraft, they remained throughout the whole war immune from any vigorous systematic effort to destroy them. Right up to the autumn of 1918 Germany was permitted to raise, smelt, and ship millions of tons of Briey ore to the Ruhr whence it returned in the form of Krupp guns or shells.

The question as to how this was possible affects one of the darkest chapters of the history of the World War. It is equivalent to asking for what purpose millions of soldiers of all nations who lost their lives after 1917 were sacrificed.

The French Chamber subsequently concerned itself with the astonishing "negligence" of the French Army Command and investigations of the circumstances brought to light irrefutable evidence that repeated attempts had already been made to deal with those responsible for it. All these attempts had, however, proved fruitless. General Sarrail, who prepared a plan for the capture or destruction of the industrial zone as early as 1914, was promptly replaced by General Gerard. Was this quite accidental? Doubtless it was, just like the refusal to adopt the suggestions made by the Deputy Engerand in the following year by an expert of the General Staff—who was a member of the Comité des Forges! In 1916, according to an admission made by the Minister of Munitions Thomas, the Cabinet themselves were seriously concerned over the forbearance displayed to Briey, when the Minister of War informed them that his orders to bomb the place had "not been carried out. . . ." Further attempts made by the Deputies Flandin and Eynac proved equally unavailing, and even when Henri Bérenger made a report on the Briey scandal to the Army Committee of the Senate, he was violently attacked by the most influential organs of the press, including *Le Temps*.

What was it that was going on behind the scenes? What secret power was paralyzing the hands of patriotic Frenchmen who were genuinely concerned for the safety of their country? This was not to be revealed until 1919. While the war was raging, few would guess the truth, and even they were scarcely likely to understand that the immunity of Briey constituted merely one phase of the mysterious business. In the course of one of the debates on the matter in 1919 we heard the following illuminating disclosures.

"If it was so easy to achieve the desired results by bombing," replied one of the owners of the Briey works, Mr. de Wendel, with a wink, "why did the Germans, who were perfectly well aware of our acute shortage of coal, not destroy the pitheads of the Pas-de-Calais collieries? . . . They were not seventeen to twenty miles distant from the front line like Briey, but only ten to twelve miles away."

The situation in the Pas-de-Calais was identical with that in the Briey sector. Tempting collieries and works, the smoke of which the German soldiers could, on a clear day, see with the naked eye, remained intact. Was that also accidental or was it, perhaps, due to a shortage of bombers? The trifling distance of ten miles and the intense activity of the Germans in adjacent sectors, render such excuses futile.

What then was stopping German and French aeroplanes from aiming a decisive blow at the hostile centres of production of war materials? In the course of the

Chamber debates in 1919 Deputy Barthé summarized the results of the inquiry:—

"I assert that the order to spare industrial establishments exploited by the enemy in the Briey region from bombing, emanated from our military authorities. I assert that our pilots received instructions to spare blast furnaces from which the enemy obtained steel and that a general who proposed to do otherwise was reprimanded." This statement is amplified by Deputy Major de Grandmaison: "This prohibition appears to have been due to an agreement between the opposing combatants. It looks as if we had told the Germans that we would not bomb Briey where they got their iron ore, if they, on their part, would refrain from attacking Bruay and the pitheads of Pas-de-Calais."

A fantastic thought! With millions of soldiers arrayed against one another in a life and death struggle, subterranean influences were prohibiting action that might bring the war to a speedy close. It is an open question whether this monstrous outrage was definitely the result of an "agreement;" more probably it was, in practice, a tacit understanding which did not compromise either party. But the undisturbed continuation of production in Briey and in the Pas-de-Calais still constitutes glaring and irrefutable proof of guilt. As the newspaper *Matin* sadly expressed the issue of the debate on the question: "War is a matter of conventions. . . ." That is just what it was; Krupp got their iron ore, Schneider-Creusot got their coal, and the war went on. Another two years and several more millions were dead. They had to die because the manufacturers in the Ruhr and Rhone valleys were not deprived of their sources of supply.

In the "Hindenburg Programme" of 1917 beleaguered

Germany mustered all her strength to take up the race against time, which was on the side of her enemies. Feverish haste was displayed in preparing war stores for offensives on a vast scale, to force a decision. North-west of Essen, beyond the old factory district, Krupp had erected immense new works with acres of covered buildings, including the largest workshop building in the world, the "Hindenburg shop" with an area of 80,000 square yards. The Krupp establishment now employed 150,000 people, working day and night, to keep pace with the output of their machines. And from down below, in the depths, came the first murmurs of resentment against war and starvation.

The flames of patriotic enthusiasm burned as brightly as ever in the Essen Head Office. How could it be otherwise as long as the entire wealth of the country was being poured down the maw of the great industrialists? In his memoirs Ludendorff admits that "money no longer played any part." The exploitation of the country surpassed all previous records and it was established by the Reichstag Industrial Committee late in 1917, that most of the supplies of war material were overcharged to the amount of hundreds of millions. The cost of steel for projectiles alone involved the Treasury in an excess charge of 500 millions. Here it was the monopoly enjoyed by the Steel Makers' Association, whose influential membership included Krupp, and who extorted all they could squeeze out of the State. This cynical profiteering constituted an obviously treasonable act, based as it was on an artificial restriction of the output of German-made steel. "No single class of contract relating to the supply of war materials was omitted or was to be omitted from the systematic swindling of the State" was the verdict of

Photo from European

SMELTING FURNACE IN THE NEW KRUPP STAMPING MILL
IN ESSEN-BORBECK

THE MIGHTY GUN WHICH BOMBARDED PARIS.
(Designed by Professor Rausenberger)

the usually pro-military Representative Noske, as the result of the Parliamentary Inquiry.

Slight opposition was offered by Berlin. All executive posts in the various Government departments and committees charged with the control of prices were held by the great industrialists themselves or by their creatures. When the force of public opinion compelled the "Iron Section" of the Raw Materials of War Department to call for a justification of the high prices charged, it encountered a wall of opposition. The axiom "price corresponding to demand" was applied even to the hard-pressed Mother Country in its hour of need, although it is obvious that this could not be put forward as justification. Noske went on to say: "Numerous undertakings, including, of course, Krupp, the A.E.G., and others, did all they could to avoid an examination of their accounts. The utmost they would agree to do was to furnish particulars, both incomplete and inaccurate, of their manufacturing costs." Whereupon General Couvette, as representative of the Ministry of War, declared the allegation to be unfounded in so far as it applied to the A.E.G. No denial, however, was given to the charge made against Krupp.

In regard to this matter also the Krupp administration preferred to await control of the press before making any statement concerning the "special price considerations" applied during the Great War. But in 1917 they did their utmost to avoid any inquiry into the matter when the question of an investigation was actually mooted. One of the senior officials of the "Iron Section" who displayed too much zeal in pressing for an inquiry was dismissed and replaced by a director of a coal and iron company, with which Krupp were on friendly terms. But a more serious campaign was launched against General Gröner, Director of the War Office, when in June, 1917, he issued a memorandum on prices, supplies, and raw materials which revealed many awkward facts, following his refusal to modify the "Auxiliary Service Act," denounced by the great industrialists as "Socialistic." For the first time in many years an imperial general had dared to oppose the wishes of the steel and coal magnates, and it was not to be endured. A secret conference of West German industrialists convened at Düsseldorf, Lieut.-Colonel von Bauer attending on behalf of Grand Headquarters. The Managing Director of Krupp and the heads of the Colliery Association violently attacked General Gröner, although they discreetly refrained from mentioning his greatest misdeed—that he had reduced the war contractors' extortionate accounts by something like fifty millions a month. The Düsseldorf "Star Chamber" proceedings soon led to the General's fall. And when in November, 1918, an investigation into the correspondence relating to the supply of war materials became possible, it had disappeared.

Krupp finished the year with over 100 millions in gross profits. This represented a slight reduction on the preceding year's trading, but was not due to any decrease in the earnings, rather to the greater skill displayed in writing down and otherwise faking the accounts. As shareholders and administration were identical, no protest was made as to disposal of profits. But the reduction of the total amount distributed in dividends to twenty-five millions, indicates that the House of Krupp reaped the greatest profit from the financial hysteria of the Hindenburg Programme. We know now that from this programme dates the financial bankruptcy of Germany and the beginning of inflation.

It was now 1918, the fourth year which saw thirty suffering and exhausted nations still engaged in the sanguinary struggle. The hearts of the combatants no longer glowed with the ardour of their original enthusiasm, but the fires in the furnaces of the great arms factories glowed as murderously as ever. As the machine dominates the worker, so the material of war dominates the soldier and the technical science of war grips and enslaves the hero. The molten metal flowing out of the crucibles and blast furnaces was actually the life blood of the nation;—steel—hard, silver-grey steel.

In this last great round just opening, Putiloff had already gone under; "only" Vickers-Armstrong, Schneider, and Bethlehem Steel were still arrayed against Krupp. Did the Essen firm really believe that they had even an outside chance in this unequal combat? With a strange lack of foresight they persisted in carrying the war to the verge of disaster by their blind belief in a "victorious ending."

As late as January, 1918, the Krupp Director Haux announced the form which this was to take. In the west the frontier was to be extended to bring the whole of Belgium under German domination; in the east, Poland and the Baltic Provinces up to the Island Oesel were to continue under German rule. He then proceeded to say: "We Germans are a nation of almost one hundred million people and occupy an extremely difficult position in the centre of Europe. We can reasonably and rightly demand that the lesser races on our frontiers recognize our situation. Such demand is both proper and justifiable and the rights of the lesser peoples must give way to those of the greater, for reasons that are morally undisputable."

These words were not spoken by Hitler, but by Krupp. The programme for German hegemony over Europe is the last word of the Essen firm. From the roots of the Pan-German Association and Defence Association, subsidized by them, grew the annexionist "Fatherland Party" of Admiral Tirpitz, whose foremost propagandist, General Keim, had to admit that he received money "for patriotic purposes" from Krupp. The weakest part of the front presented by the Central Powers was Austria, whose Foreign Minister Czernin was suspected of pacificism; so Krupp proceeded to incite their friends in thought, the Hungarian press, against him. Vienna newspapers assert that no less than two million marks of Krupp money were distributed among the various Budapest journals for that purpose. The fight against the slack attitude of their own country was prosecuted with German thoroughness, and in order to influence the most prominent anti-war agitators Krupp actually caused their creatures to organize an "Association of War Wounded" in order to win these unfortunate beings to their cause. When the outraged members discovered how they had been duped, accounts of the association were, in accordance with the time-honoured practice in such cases, found to have disappeared.

On March 23rd, 1918, General Ludendorff began a last offensive, the Second Battle of the Marne, the greatest battle of the world's history as regards materials of war.

In the Laon Salient, near Crépy, where the front line was at its minimum distance from the French capital, stood the House of Krupp's latest contribution to Germany's decisive offensive—a gun which bombarded Paris at a range of over 80 miles. This was a technical sensation even in the fourth year of the War, and before it was realized in practice it might have been looked upon as an impossibility. How had Krupp managed to achieve the impossible?

By means of a monster, a technical monstrosity! The gun barrel was 112 feet long, with a diameter of 39 inches, and a bore of 8½ inches, so that its walls were 15¼ inches thick. The concrete bed weighed 60 tons, the gun mounting 50 tons and the gun barrel 40 tons. This huge gun, weighing in all 150 tons, was mounted in the middle of a dense wood, skillfully camouflaged from hostile aeroplanes and closely guarded by a military cordon.

All the basic principles of gunnery were upset, meaningless. All the projectiles were carefully numbered, their diameter increased from 8.2 to 8.4 inches, and their weight from 200 to 230 pounds. After every round the gun barrel was slung from special blocks and had to be "straightened." Every second projectile was larger and longer. After 65 rounds the gun barrel had to be replaced by a new one.

Firing at a range of over eighty miles was a brilliant achievement of ballistics and was based on complicated calculations which not only concerned direction and elevation but also atmospheric pressure, force of wind, weather, and even the curvature of the earth's surface. The projectiles required propelling charges of six hundredweight of explosive per round, comprising two varieties of propellants which had to be "preheated" in an underground magazine. The gun was fired at an elevation of fifty-two degrees and the projectile took one and a half minutes to travel through its trajectory, penetrating into the stratosphere to a height of 26 miles and a temperature of three degrees F. to reach its target.

The gun's crew numbered sixty seamen gunners, of whom thirty were required to man the blocks and tackles for straightening the gun after firing. A full Admiral was in command. He was in direct communication with Grand Headquarters by a special telephone cable and was also directly connected with thirty batteries located in the neighbourhood, with aerodromes and meteorological stations.

Each round was accompanied by a simultaneous salvo from ninety guns of adjacent batteries in order to render the task of enemy flash- and sound-detecting detachments more difficult. Forty aeroplanes guarded the sky over Crépy, and a chain of spies, distributed from Paris to Switzerland, served to collect and transmit reports on the hits made.

Each shot cost 35,000 marks.

From March 23rd to August 9th, 1918, this monster gun fired a shell at Paris every twenty minutes; 320 reached the city, and of these 180 burst in the centre and 140 in the various inner suburbs. They caused the loss of over 1000 human lives and destroyed much valuable property. On Good Friday a shell burst in the church of Saint Gervais, killing ninety-one and wounding over 100 persons. This might have continued for years if the Allied advance in July, 1918, had not put an end to the terror. The dream of the Cannon King, to win victory for Germany by a bombardment of Paris, had not come to pass.

The destinies of Germany in the autumn of 1918 were governed not only by important military and political factors but to a considerable extent by the complex of the monster gun and the House of Krupp generally. The long-range gun which bombarded Paris represented the peak of Essen's historic achievement—unprecedented mastery of materials, uncanny precision, bold perfection of metallurgical, constructional, and ballistic inventiveness. Nothing more! The "more" is intuition and intuition was expressed in the word "tank." The tank did not originate in Essen. Generations of unenterprising monopoly resulted in failure at the critical moment. What was merely the ill-luck of a private firm when the Gruson armour and the recoil cylinder gun were misjudged, became the tragic fate of Germany in the case of the tank. When the steel war chariots broke into the trenches of Cambrai, the retreat of the imperial armies began and their flight was headed by a ghostly symbol of Essen's shortcomings—the defeated monster gun.

At the end of August, Ludendorff suggested that the Emperor's eyes should be opened, by a special delegation. Krupp von Bohlen, Stinnes, Duisberg, and Ballin were to explain the fatal gravity of the situation to him. But nothing came of it; doubts expressed by court officials found these great industrialists only too ready to listen. It was enough for these gentlemen that they should have been afforded an opportunity of making money out of the war and they were prepared to leave the thankless and painful task of liquidation to others.

Things now moved quickly. Early in September William II, filled with apprehensive forebodings, decided on a personal attempt to induce his exhausted people to carry on. His choice fell on Essen. A strange coincidence that the Empire, the earliest beginnings of which were founded on a violation of the Constitution instigated by Krupp, the thunder of whose guns and the protection of whose armour helped it to become a World Power, should in its last hour turn to Essen with its appeal of desperation!

It was, indeed, an appeal of desperation which the Emperor attempted to make on that 9th of September, 1918, as spurred on by terror he hurried through "Gate No. 28" into the huge works, ignoring the expostulations of his retinue. He strode through the noisy, sooty iron foundry building and for the first time in his life faced an unselected and unrehearsed assembly of his people.

Fifteen hundred Krupp foundrymen, formerly uncritical works employees, but now rebellious from four years of starvation and disappointment, were the unwilling audience for this last speech of the German Emperor.

It was quite useless for the grey-uniformed bemedalled monarch to address the distrustful, almost hostile men clad in war-time shirts of paper and clumsy wooden clogs as "my friends," to remind them of the mourning and sorrow which were the common lot of the whole nation, sparing neither the princely mansion, nor the humble workman's dwelling. He had no answer for the question which rose to the lips of all his listeners and which he could read in their faces—"When shall we have peace?"—other than the worn out platitudes of the patriotic press. The Emperor's A.D.C. reported later: "Faces became grim, and the more excited the Emperor grew, the more clearly was the negative attitude of his hearers expressed."

The speaker was now quite in his element: "It is now a question of making the supreme effort on which the whole issue depends. Our enemies are well aware of it, because they have the utmost respect for the German army and as they realize that they cannot overcome our army and our navy, they are endeavouring to defeat us by spreading false reports and making mischief within our country, causing internal disaffection. This is not due to any action of the people but is artificially fostered. Anyone who listens to such rumours and spreads unconfirmed reports in railway trains, factories, or elsewhere, is committing a crime against his country and is a traitor deserving severe punishment, be he count or workman. I know quite well that each of you agrees with me."

Mistaking the stony silence of his audience for acquiescence, the Emperor thought that a last appeal would suffice to regain their former confidence.

"Be strong as steel and the unity of the German people, welded into a single block of steel, shall show the enemy its strength. Those among you who have decided to act on this appeal, those whose hearts are in the right place and who will keep faith, let them stand up and promise me in the name of all the workers of Germany: We will fight and hold out to the last man. So help me God! Let those who will do this, answer me —Yes!"

The official account of the address states that the reply was . . . "loud and prolonged cries of Yes!"

In actual fact a few isolated and hoarse affirmatives were given by the nearest foremen and engineers, interspersed with hollow cries of "Hunger."

The Emperor heard them and grew pale. But a moment later he tried to save the situation by simulating deafness—

"I thank you. With this 'Yes' I will go to the Field-Marshal. Every doubt must be swept from heart and mind. God help us. Amen. And now, men, farewell!"

Badly shaken, William left Essen. A few hours later a whispered rumour of an attempt on his life in the Krupp works ran like wildfire round the industrial district. Even when it proved false and unconfirmed, the impression that the spell had now been broken grew stronger amongst the openly disgruntled workers.

Barely sixty days after the Imperial speech and the hunger cry of the Krupp foundrymen, the Revolution broke out.

Before leaving these great days, there should be a reply to the question so often asked: what did Krupp make out of the war? No accurate data were ever allowed to come out, as the figures quoted in the published accounts of the firm were subjected to drastic reduction. But it will suffice to refer to what the firm themselves admit.

There were first the figures termed "gross profits" which might actually be considered almost net, as new buildings and reconstruction were paid for out of current receipts and their cost was, therefore, already deducted from gross profits. The latter appear as follows:—

1914-15 - - - -	128.2 millions
1915-16 - - - -	143.3 millions
1916-17 - - - -	103.8 millions
1917-18 - - - -	56.9 millions

Although the last return was made after the general collapse and may, therefore, be regarded with some suspicion, the grand total amounts to the respectable sum of 432 million marks of gross profits, not counting the huge expenditure on new construction! 225 millions more than the total earnings in the previous four years of peace, which of themselves were years of unprecedented boom.

Compared to these figures, the alleged net profits of 226 millions are merely an indication of juggled accounts and secret allotments to reserve. Even so, these net profits are 100 millions greater than those shown by the record earnings of the previous four years of peace.

What then did Krupp really make out of the war? They admit 226 or even 432 millions, but say nothing of the value of the new buildings and modern plant which they acquired at the same time. The figure of 800 millions given to the author by a well-informed person in Essen is substantially nearer the truth. This is equivalent to four and a half times the share capital at the outbreak of the war and corresponds to the earnings of twenty good peace years, easily come by in a short war period.

This accurate estimate of the millions which poured into Essen is an actual fact not always realized outside Germany. Herr Krupp von Bohlen has had the audacity (wisely enough, not until January, 1934) to add a new one to the numerous legends woven around the history of the Essen firm, to the effect that Krupp declined to make any profit out of the war. At the outset of the Hitler period of rearmament he announced: "It was a basic principle, dating from the day the War broke out, that the owners of the business had no desire to make money out of it."

An amazing statement! Has the world hitherto failed to recognize this heroic sacrifice of hundreds of millions? If the bold statement is carefully analyzed and due weight given the expression "owners," the matter appears in a somewhat different light; it remains on record that Krupp was shrewd enough not to allow the annual dividends distributed during the war to exceed twenty-five to thirty million marks. As to the surplus profits, of which only a modest proportion were devoted to the victims of the war, the "owners" waived their claim in favour of—the firm! The one is identical with the other, therefore the alleged sacrifice was merely an internal transfer of profits equivalent to changing them from the right-hand pocket to the left-hand one.

THE CONVERSION

On November 9th, 1918, work in the Krupp establishments ceased with an audible jerk. For the first time in a century the hammers were silent and the army of 165,000 workpeople left the paralyzed workshops. Was it the end?

It almost seemed like it. The stranglehold on Essen was not merely the suspension of arms production, which had kept Krupp alive for the past four years; output could be altered to embrace other classes of products, even if the temporary chaos due to demobilization rendered such schemes wildly speculative at the moment. The real danger that threatened Krupp came from another quarter. What would be the decision of the victorious Allies, who must obviously regard the destruction of the armament centre of Essen as a seal on the defeat of Germany? There was also Berlin to be reckoned with. The Socialists were in power there, the avowed enemies of the firm in a conflict extending over a lifetime. The nationalization of key industries, more especially of the arms industry, was always an important feature of their programme, which they were now in a position to carry through.

Whence would the death blow to the House of Krupp come, from Paris or from Berlin?

Ballin, the great shipowner and friend of the Emperor, had sought refuge from similar uncertain dangers in death. Kirdorf, Fritz Thyssen, and others were intriguing and planning to maintain their rights of ownership, with longing looks over the Rhine where the foreign armies appeared to offer a refuge from the revolutionary onslaught. Only Hugo Stinnes, clear-headed accountant and student of human nature, exerted himself to avert the threatened calamity; he had just succeeded in reconciling the trade unions with the employers in a Working Commonwealth, which was later destined to be termed "a bulwark of defence against the quagmire of revolution."

What was Krupp doing? Hugenberg's recent resignation enabled Gustav von Bohlen to assume undisputed control of the Krupp concern. A man of average ability, he was a rigid upholder of the interests which he had acquired through marriage, and behind an appearance of silent reserve he concealed a cautious singleness of purpose allied with ruthlessness. His policy was now cautiously to feel his way amongst his altered surroundings and to establish new contacts with the new rulers, without, however, breaking with the old ones.

The first thing was to overcome the revolutionary unrest amongst the employees of the vast concern, which took the form of aimless strikes, demonstrations, and a violent attack on the Rheinhausen establishment. A conciliatory attitude had to be adopted, coupled with threats of bankruptcy which would affect everybody. The

Board of Directors issued a solemn announcement to the effect that "the share capital is being gradually dissipated" and that "soon all that is left of the works will be a heap of ruins." Coming just after the rich harvest of the war years, this was a pretty bold attempt to trust to the shortness of human memory, but it proved effective. The reins were slowly tightened as the position was steadily consolidated. An endeavour to incite the long-service personnel against "the newcomers" amongst the employees proved effective. Within a few months sixty per cent. of the 108,000 male and female employees were dismissed and the remaining 43,000, alarmed and divided among themselves, were much more amenable.

In Berlin the position was clarified with amazing rapidity. The steel magnates, drawn even closer together since the events of November, whispered confidentially that the official Socialist programme was mere eyewash designed to keep the multitude quiet. A congress of all armament workers convened at Erfurt to discuss the nationalization of the arms factories was, accordingly, not taken seriously. Among the many fateful pacts concluded by the young Republic with the old Powers, its capitulation to big business was the most far-reaching. Exactly four weeks after the Revolution, the Demobilization Department issued a secret decree, which could scarcely have been more pro-capitalist under the old Imperial régime; it authorized the armament firms to receive payment for war stores contracted for, but undelivered. As the Socialist newspaper *Vorwärts* described the case in a legal dissertation—a *damnum emergens*—a loss that had been suffered; in plain language, compensation for loss due to a premature ending of the war.

Instead of nationalizing, the new régime began by subsidizing.

In the general confusion of the collapse this opened wide the door to a wholesale plundering of the State coffers. War profits were finished, long live post-war earnings! The Daimler Motor Company, exposed as shameless profiteers in the spring of 1918 and then threatening a production strike, which led to the works being put under military control by Ludendorff, were now presented with a deferred payment of—sixty millions! It should be noted that this payment was *not* in respect of any particular deliveries. The sum collected by Krupp for undelivered war contracts was substantially higher, although the actual amount has not been disclosed. In matters relating to big industrial undertakings, the Republic displayed a truly chivalrous discretion.

But the victorious Powers? Would not the hatred of the enemy enable them to devise effective means of destroying the ever-present danger spot in Essen? Such fears were speedily dispelled. The statesmen and experts who had met in Paris to decide the future of the German arms industry did not go beyond the mere destruction of machinery and buildings which, in view of the vast material wastage incidental to four years of war, was not of itself a really fatal blow.

Under the demands put forward by the Inter-Allied Commission of Control, Krupp were required to destroy

159 experimental guns
379 installations (cranes, furnaces, reservoirs, &c.)
9,300 machines of all kinds.
157,000 cubic yards of concrete and earthworks
801,420 tools and appliances.

These figures appear enormous, but in reality they represent little more than the normal wastage of the next few years in any really large undertaking due to the march of technical progress and rationalization. The main difference lies in the fact that the destruction in the case of Krupp was paid for by the Government! It took a long time for the firm to assess their total claims for compensation in respect of the property destroyed, but they were finally computed at 103 million marks. An agreement on the matter was arrived at in Berlin on April 23rd, 1925, and, after allowing for various payments made on account, a final sum of eleven millions was paid over in settlement.

A second condition imposed by the victorious Powers forbade Krupps to make arms. All they were permitted to manufacture were four guns per annum and such armour plate as might be required for replacement units of the German navy, subject to the I.A.C.C's. approval of the plant used for this purpose. It is difficult to see what the originators of these restrictions thought they could effect thereby. Obviously, deliberate and open violation of the prohibitions could be prevented as when at the end of May, 1920, a surprise visit by officers of the I.A.C.C. revealed that Krupp had resumed the manufacture of 3-inch field guns, ostensibly for the Reichswehr. The restriction clauses were, however, useless for the purpose of preventing cautious circumvention by means such as the use of the machines readily adaptable for making war material. When, after a prolonged and costly period of activity, the Inter-Allied Commission of Control finally reported the completion of Germany's disarmament, Krupp had already got as far as the estab-

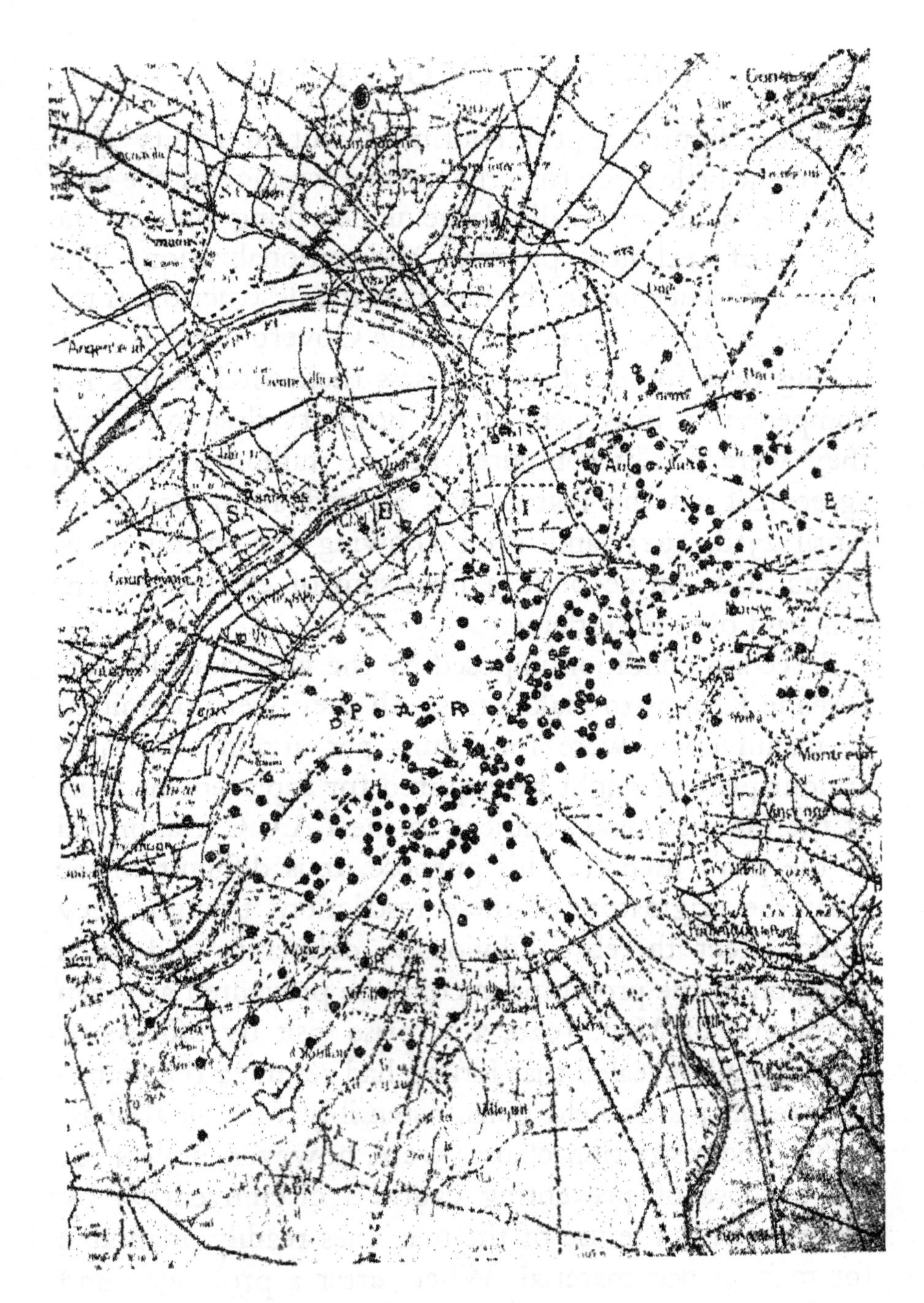

SHELL HITS IN PARIS.
(From March 23 to August 9, 1918, 320 shells fell inside the city limits.)

Photo from [illegible]

AN ELECTRIC FURNACE IN THE STEEL PLANT AT ESSEN

lishment of secret subsidiary manufacturing facilities outside the area of the Commission's supervision.

The only really effective means of preventing future production of materials of war by Krupp, sequestration of the firm's financial resources which could have been carried out in 1918-19, was conspicuously omitted.

As the dangers threatening from Berlin and Paris had proved groundless, all obstacles to the resumption of production had been removed. The plan mooted in the consternation caused by the general industrial collapse, of completely abandoning the original works in Essen in order to concentrate on the modern Rheinhausen establishment, was dropped—probably after consultation with the Berlin authorities, whose interest in the continued existence of the Krupp works was maintained even in Germany's darkest hours. In 1918 the Government had already approved a contract for the supply of 100 locomotives and about 2000 railway trucks per annum by Krupp. Official foreign trade departments were also concerned with safeguarding the firm's interests. When the first "Krupp Cash Registers" appeared on the market, the importation of a famous American make was promptly prohibited, a prohibition removed only much later. The great conversion of the armament works into a manufactory of complete ranges of special products of commercial utility, lauded as a remarkable technical achievement of the firm, would have been impossible without the secret backing of the State and its financial assistance. Krupp's privileged position as a "national institution"—as it was once formally proclaimed by the old Emperor William I—had not been assailed even by the events of November, 1918.

Before a new programme of production could be put

into effect serious difficulties connected with the supply of raw materials had to be overcome. Although the loss of iron-ore supplies in Lorraine did not amount to much in the case of Krupp, they were badly hit by the loss of the Minette. Their holdings in the Orconera Iron Company, the Spanish ore concern in Bilbao and another subsidiary undertaking in Riga, were forfeited during the war and it was now a case of seeking new sources of supply. This was provided for by the acquisition of two iron-ore concerns in Finland and by the conclusion of large scale purchase agreements in Sweden, shipment being undertaken by the Devon Erts Maatschappij of Rotterdam, in which Krupp were now interested. There were also huge mountains of scrap in the various Krupp establishments arising from destroyed or unfinished plant and products. A new process enabled scrap to be utilized for the manufacture of certain grades of steel. When a shortage of coal was experienced and the State quota allowed for Essen proving inadequate, Krupp had no scruples about arranging with the Hamburg firm of Blumenfeld to import British coal.

The great conversion from war- to peace-time products, carried out from 1919 to 1921, turned the ordnance works into an industrial warehouse. Everything could be had there, from heavy locomotives to artificial teeth (of rustless steel and guaranteed tasteless), from ordinary black plates to cinema projectors, from typewriters and cash registers to water sprayers and ships' propellers, from dairy machinery to tractors, dredges, and trucks, from cameras to textile machinery, from agricultural implements to calculating machines. A varied range of products without any special lines, the slogan being: "We make everything!" Wherever the directors saw an opening for a product in German industry, or where competition was weak, they made a dead set at it. Everything was of a more or less improvised nature, so that for years after the conversion the firm's activities still appeared temporary and transitional. This was further emphasized by the care taken in regard to some of the alterations—with an eye to the future; the former armour-plate rolling plant turned out the thickest plates in the world, the lorries delivered to the Government and to local authorities had chassis which were readily adaptable for artillery purposes, and the tractors were obvious modifications of light tanks. Viewed in the light of the ultra-rapid rearmament of Germany after 1933, these early measures may seem very modest, but they paved the way to secret developments and agreements of a more permanent nature, the psychological effect of which was to go far in creating the right atmosphere for future treaty violations.

Contemporary problems were, however, of more immediate urgency. Having ensured the continued existence of the works, the next question was how to safeguard the proprietary rights of the Krupp family, which appeared to be threatened. Shortly before the national collapse, the Essen firm had recourse to Dutch currency to save themselves and had received about 100 millions on the security of a block of their own shares. Their redemption was, of course, provided for and duly effected at the first opportunity. This exclusive ownership, frequently so awkward, was not wholly due to family pride; executives of the Essen Head Office were fully alive to the fact that the future well-being of the firm would continue to depend on the existence of a single and perfectly discreet proprietor who could deal with the State.

From 1918 to 1921 no dividends were paid by the firm and the balance sheet of the first post-War year showed a heavy loss of thirty-six millions. These figures, however, disclose little of the real situation, for while an appearance of poverty was simulated outwardly, a process of internal consolidation, carried out at a surprising rate, was going on. The greatest drag on the firm's resources, the now superfluous Munich works of the Bayrische Geschutzwerke, were sold at a good price to the Nuremberg firm A. G. Neumeyer, a subsidiary of the Haniel concern. In order to avoid patent difficulties arising out of the conversion, reciprocal licence agreements were concluded with the various special firms concerned—as with the Baden Engineering Works of Fahr, in regard to agricultural machinery, and with Ernemann-Dresden for photographic cameras. In each case the Essen firm, as the greater and wealthier party, got much the best of the agreement. In December, 1920, the first of a series of new purchases took place; 500 acres of industrial building land were bought in Merseburg, near the Central German lignite mines, where the new industries of the post-War era were most likely to be developed. This transaction was followed by a rapid succession of purchase agreements, carried through in the traditional Krupp manner, with the Konstantin and Helene Amalie collieries, which ensured a coal supply of ten million tons. Finally, the association with the former Ehrhardt Rheinmetall Works was put on a firm basis; here Krupp was associated with Dr. Rathenau's A.E.G. and with Messrs. Otto Wolff of Cologne, whose connection with Krupp had turned them into iron and steel merchants in a large way of business. At the end of 1921, just three years after that fateful November, Krupp were in a position to consider the setback experienced through national collapse as more or less overcome.

The cautious and conservative "foreign policy" during this period of reconstruction had been determined by the head of the firm, Krupp von Bohlen. The creative impulse was, however, lacking in him and was provided by a comparative newcomer, who had only recently become prominent—Dr. Otto Wiedfeldt, the commercial manager of the firm. His career is that of a typical Krupp director, with its indefinable mixture of political and business achievements. After being a City Representative for Essen, where he first came into contact with the firm, he quickly rose to an important post in the Ministry of the Interior, but relinquished this to become Adviser to the Ministry of Transport in Tokio and to the Administration of Korea, where Krupp had important railway and mining interests. During the war he worked under the Food Department, which again brought him into close contact with Essen, Krupp being associated with the Reichsgetreidegesellschaft (National Cereal Corporation). In this capacity he pressed for increase of agrarian prices. After a most interesting period of service as Head of the German Ukraine Delegation, Wiedfeldt openly entered the employ of Krupp, where he was largely responsible for carrying through the technical conversion of the works. He also represented the firm in Versailles, Spa, and Weimar, where, less obtrusive than Stinnes, he was one of those influential personalities in the background, whose names rarely figured in lists of official delegates. It is interesting to note that in 1922, just when German industry was establishing financial contact with the United States, Wiedfeldt "left" the firm to become German Ambassador in Washington (during which time Krupp succeeded in negotiating two important American loans, which rendered it possible to make the adjustments involved by a stabilization of currency), and in 1926, shortly before his death, he rejoined the Board of Krupp. His career presents a typical picture of the life of a Krupp director. They were no Zaharoffs, these gentlemen; their origin was not shrouded in mystery, and their only knowledge of prisons in London and Athens was derived from sensational novels. They were respectable citizens, honourable men of business, and conscientious officials—up to the point where they contacted with that semi-darkness between politics and big business and where, in their own way, they proved themselves to be quite the equals of the adventurer from Tatavla.

Wiedfeldt's outstanding personal achievement was the establishment of contacts with Moscow. Although it had ever been one of the oldest traditions of the House of Krupp to look to the East, the connection between this conservative firm and the land of world revolution was none the less surprising. The first of these transactions had its unusual beginning with a prologue in Stockholm, where in the summer of 1920 conversations took place between the Soviet Delegate Krassin and the leader of the German Socialist Trade Unions, Legien. The German trade-union officials of the A.D.C.B. had founded an Economic Corporation with the assistance of Swedish credits that they might secure a commission on all business with Russia. The scheme fell through when Wiedfeldt began private negotiations with the Russians. The

Moscow authorities placed contracts for locomotives to the value of 500 millions in Essen, these contracts being distributed among the members of a syndicate headed by Krupp.

The advantages to be derived through close contact with Essen by isolated Russia, ravaged by war and civil war and formerly one of Krupp's best customers, were fairly obvious. But what were the aims of the Essen firm? The agreement concluded at Rapallo and the attitude of the Reichswehr accounted for much, but the next great transaction did not altogether fit in with these considerations. It concerned an "agricultural concession" of some 247,000 acres of the richest land available for agricultural purposes, and apart from the earnings to be anticipated from harvests with modern machinery, Krupp was said to be calculating the potential extension of their exports of agricultural machinery to the land of ten million peasant holdings. But did this explain it all? Was not Otto Wiedfeldt, initiator of this somewhat unusual business for an industrial firm to handle, one of the exponents of the Ukraine schemes of Imperial Germany? Was German neo-imperialism desirous of "dropping a card" on Eastern Europe, as an indication that it wished to be included in the party, if there was "anything doing"? this seems to be borne out by the fact that, along with Krupp and Otto Wolff, about seventy-five per cent. of British capital was involved, even if the importance of speculative undertakings of this nature tended to diminish as Russia's internal consolidation increased. For the rest Krupp got little joy out of the Donetz concession and disputes as to the interpretation of the agreement soon arose.

It was now 1922—just before the inflation and occupation of the Ruhr. The year also marked the completion of the consolidation of Krupp and of German industry in general. An abyss yawned behind the steel and coal magnates and they shuddered over it in their meetings of the Ruhrlade Club in Hügel. The spectre of revolution again threatened them, when in December, 1920, the National Assembly adopted the "Sacrifice for State Necessities" Act, under which all incomes above 5000 marks were subjected to a capital levy up to sixty-five per cent. and which would not allow any German citizen to own a fortune of over 370,000 marks. There was, however, no need for alarm, nothing was going to happen to anybody—except to the creator of this law, Erzberger, who was soon rendered harmless by gunmen's bullets—just as, some years earlier, it had happened to Karl Liebknecht, detested for so long and already menaced with "bloody vengeance" when accusing Krupp at the time of the Brandt trial. Still the patient Republic, so frequently misled and betrayed by its reactionary bureaucracy, had dared to kick against the capitalistic pricks! The answer was, as we have already said, inflation and trouble in the Ruhr.

The part played by Krupp in the currency manœuvres *à la* Stinnes, is difficult to prove, as their past experiences in armament transactions had taught them, far more thoroughly than they had taught Stinnes, how to avoid publicity in such little matters. The attack on the mark by the Federation of German Industries was headed by its President, the Krupp-Director Sorge, and his firm were also engaged in hoarding. They made use of Holland as depository for their valuable foreign capital, having established useful "sidings" there during Germany's difficult and uncertain years. The whole of the Krupp overseas trade went through the Allgemeene Overzee'-sche Handelmaatschappij in The Hague and the profits were concealed in private accounts in the Amsterdam-sche Credit Maatschappij, one of whose directors was the Krupp Director Buschfeld. It is only possible to guess at the millions lost to the Reichsbank through this channel and of the use to which they were put in attacking the mark.

The truth of these statements is amply proved by the expansion of Krupp during these years. If Stinnes's noisy activities at the peak of his inflationist machinations are carefully traced, it is found that Krupp was invariably a good second to him. Thus, at the time of the purchase of one-third of the share capital of the Berliner Handelsgesellschaft (Berlin Commercial Corporation) which caused such a sensation in Germany in October, 1922, the fact that Krupp had, at the same time, increased their holdings in this highly reputable German banking concern, was lost sight of. The first stage of the Stinnes movement towards South-East Europe, the acquisition of the Böhler A.-G. in Berlin, with ore mines, ironworks and steelworks in Styria, Lower Austria, and Bohemia, revealed that Krupp had already secured possession of a substantial holding in that company. It was the same with the developments in the explosives industry, where Krupp, starting from Köln-Rottweil, made a vigorous thrust to expand. The attention aroused throughout the world by ambitious purchases made by Stinnes, completely overshadowed the connections between Essen and the Spanish Levant Shipping Union, the Brazilian Monteney Coal Company, and the magnesite mines in Styria, Chile, South Africa, and Turkey. The main difference between the procedure adopted by

Stinnes and Krupp in this policy of expansion based on the tottering mark, was that the latter felt their way far more cautiously and were chiefly concerned with the building up of their own business, Krupp also gambled in blocks of shares which controlled the destinies of tens of thousands of people; they disposed of their holdings in the Mannesmann A.-G., dating from the days of the Moroccan venture, to Dutch interests, characteristically omitting to notify the German undertaking concerned.

At the end of 1922 the Krupp family sprang a pleasant surprise on their army of workers by announcing that they proposed to assign an interest in the business to their works employees. It was not a gift they offered, but an opportunity to acquire 100,000 shares at 1000 marks per share, with a minimum guaranteed interest of six per cent. A compliant press hailed this announcement with exultation and praised the "practical Socialism" of the magnate of Hügel, who was ready to share his property with his employees. When details of the scheme became known, it appeared that this was a gross exaggeration. With a capital of 500 million (depreciated) marks, 100 million were still only a fraction, the holders of which could not expect to exercise any influence whatever, let alone possess any measure of control, over the firm's policy. Any such possibility was guarded against by a skillfully worded regulation under which employee shareholders were debarred from personal attendance at general meetings (a privilege to which the humblest shareholders are normally entitled), but had to form representative groups with corporate representation. Share holdings ensured an overwhelming majority for the firm's official representatives in such an assembly, so that the voting power of the employee shareholders was a polite fiction. Care was taken that the participation of the employees should not become too intensive, nor likely to exceed the amount of their savings.

The Board of Directors turned down the scheme and up to the end of 1922 only thirty-two shares out of 100,000 were taken up, when growing inflation put an end to the whole thing.

The final collapse of the mark brought a sequel to which histories of the firm of Krupp are ashamed to refer: the disappearance of the pensions funds and endowments to which so much publicity had been given. Money in these funds was partly invested outside, but thirty millions of gold marks were, in 1914, invested in the Krupp A.-G. at the modest interest of four per cent., although the firm made twice that amount on the loan. All this now shrank to nothing, the insurance fund subscriptions and vaunted welfare endowments were swept away. While the firm itself had richly profited by its "welfare" publicity, a profit which made up for the cost, the Krupp pensioners were now called upon to endure poverty and hunger in their old age. The tragedy of their fate evoked much public comment for a decade. "Fidelity for fidelity" was the Works slogan when these old people offered it their labor-power, their devotion. But all that was now forgotten. Finally, the Government was compelled to intervene, although it stipulated that the firm should contribute an amount equal to that of the Government grant. Even this assistance barely sufficed to provide a starvation pittance for the unfortunate pensioners, and drove these peaceable old workmen to pitiful public demonstrations in the streets of Essen.

The theme "How Krupps provide for their old employees" was deleted from the Essen propaganda repertory after the inflation.

The entry of French troops into the Ruhr industrial district in January, 1923, marked the beginning of a period of great internal and external trials for Germany's industry. It may well be asked whether Krupp was among those whose truculent attitude provoked the step taken by President Poincaré. They certainly held aloof from the efforts made by Rathenau and even by Stinnes to come to a trade agreement with France. When faced with the problem of reconstruction after the general collapse, Krupp, Haniel, and Thyssen unanimously elected to look to England for assistance, and soon succeeded in getting into contact with British capital, by way of Holland.

When in those January days of 1923 detachments of French troops from Düsseldorf crossed the Ruhr and occupied Essen and a general appeared at Hügel as an uninvited guest, the shrewd foresight of the House of Krupp was again displayed. There was no need for any hurried and romantic flight like that of the Coal Syndicate; the secret archives of the firm and of the family, comprising documents of absorbing historical, political, and business interest covering half a century, had long since been removed to safe custody in the interior of Germany, and to provide for any contingency, a new head office for the Krupp concern had been established in Berlin, under the title "A.-G. for Undertakings of the Iron and Steel Industry." As in August, 1914, Krupp had not been found unprepared.

Just a little stirring up of the fire, and out of the universal dislike of the "bayonet," as it was called to catch the mass, flamed enthusiasm for passive resistance. On March 31st a French military patrol appeared at the Krupp works to requisition trucks from the central garage near the Head Office. While the soldiers were attempting to force the locked doors, a works alarm siren rang out, and within a few seconds hundreds of sirens were howling all over the immense area of the works. A huge crowd of workers rushed to the garages and surrounded them. The deafening noise made by the alarm signals made speech impossible, the front ranks of the men were pushed on by the pressure of those coming up behind them and the circle round the handful of alarmed soldiers was drawn tighter. The French *poilus* in their flat helmets began to look worried over the dangerous situation, in which thoughtless superiors had landed them. Trade-union officials hurried up and endeavoured to persuade the men to go back, but their efforts proved fruitless in the face of whisperings from conscienceless agents.

The position of the soldiers was becoming more serious every second. The leader of the patrol, after calling out several unheeded warnings, gave the order to fire. Thirteen dead and forty wounded lay on the flagstones and the trucks were free to depart.

Krupp von Bohlen and nine of his co-directors were arrested and tried by court-martial, receiving sentences of from fifteen to twenty years' imprisonment, with fines running into hundreds of millions of paper marks. A summary justice exercised by one guilty party on another. And the thirteen dead? Little was said about them. They were merely incidental, even if of tactical value, in this conflict over reparations and industrial quotas.

From May till November, 1923, the head of the firm

of Krupp and his co-directors had to remain in Düsseldorf Prison, but their incarceration was no martyrdom, as the French generals knew what they owed to a German industrial magnate. The cell doors of these gentlemen were allowed to remain open; they were permitted to be together; their food was provided by the Red Cross; unaccompanied walks in the garden were allowed until evening, when they all assembled—all this was seen and heard by the writer of this book in the autumn of 1923 when he spent many weary weeks of imprisonment while awaiting trial before a French court-martial in a cell almost adjoining that of Krupp and his co-directors, and meeting them at exercise in the prison yard, one side of which was formed by the outer wall of the Rheinmetallwerk—formerly owned by Ehrhardt, but then belonging to Krupp. And again legend proved stronger than actuality. The picture of the "martyr" Krupp von Bohlen in the hands of brutal soldiery is an indestructible requisite of future nationalistic propaganda. Herriot's Christmas amnesty gave Krupp freedom. According to the French scientist Gaston Raphael, Krupp bore no malice and was one of the first to sign the so-called "MICUM" agreements, in which the Ruhr magnates made their private peace with France.

Despite defeats abroad, the crops sown by the great industrialists were beginning to promise a rich harvest at home. On June 6th, 1925, Stresemann wrote in his diary: "Then we had to raise 50 millions for Krupp." The Government was now headed by Dr. Hans Luther, Burgomaster of Essen at a period when the "three class electoral system" enabled Krupp to control the majority of the municipal voting power. Dr. Luther was now in a position to requite their former benefactions in kind. He proceeded to do this without consulting the Reichstag and in violation of both constitutional and common laws, raided the Treasury and paid "compensation" to the great industrialists for damage and loss caused by the campaign in the Ruhr to the tune of 715 million gold marks. This payment was later investigated by a commission of the Reichstag, but no effective steps were taken against the guilty Chancellor. The names of the recipients of these 715 millions were, of course, not disclosed. Only one item was revealed—the payment of seventy-five millions to the firm of Krupp.

With "compensation" for cancelled war contracts, for property abroad lost, and destroyed armament material, payment for the Ruhr campaign losses made the fourth large subvention granted to Krupp. The total amount of these payments may be estimated at 300 million gold marks. No wonder then that when the political situation grew easier, the generous Republic enjoyed something like toleration from the great industrialists. Questions arising out of socialization and taxation having been adjusted, the inconvenience caused by trade-union demands was accepted as a necessary evil. As long as the new State looked after the interests of business and dealt with the consequences of the collapse in an adequate manner and, more especially, as long as it lent its assistance in the liquidation of foreign commitments, it could reasonably count on the passive loyalty of the industrial magnates.

There was little publicity concerning Krupp activities during the next five years. Their function as the armourers of Germany had ceased, and new technical developments in Essen such as the discovery of rustless steel or the improvement of new manufacturing processes can scarcely be regarded as epoch-making. After the stabilization of Germany's currency, the Krupp Company's share capital was fixed at the relatively modest amount of 160 millions. In association with Thyssen and the Phœnix Company they moved up to first place in the German raw-steel industry and to second place in the Coal Mining Syndicate. Some of the men who had achieved industrial greatness under the auspices of Krupp now surpassed their patrons in the extent of their many-sided interests; among them Otto Wolff, who had made millions out of the Rheinhausen products. After the death of the Krupp-Director Sorge, who was a member of the Reichstag, the Essen firm lost their traditional Presidency of the Federation of German Industries, which now passed into the hands of the I. G. Farben concern—an indication of exhaustion and resignation. The Krupp A.-G. appeared to be merely turning into one of the dozen large and increasingly industrial undertakings located between the Rhine and the Ems.

But not entirely. When the great Ruhr Trust was formed by Thyssen, the Phœnix, Deutsch-Luxemburg and Gelsenberg Companies, the Bochum Union, and others, under the name of the "Vereinigte Stahlwerke" ("United Steel Works"), with a share capital of 800 millions and 200,000 employees, thereby becoming the greatest industrial concern of its kind on the Continent, Krupp held aloof. The reasons for this action, which caused a good deal of public comment, do not sound convincing: "We do not consider that it would be in the interests of Germany's trade for famous names like that of Krupp to disappear. Moreover, the system of development and expansion of a family concern like Krupp's differs from that of the other iron and steel

DR. GUSTAV KRUPP VON BOHLEN UND HALBACH

BERTHA KRUPP VON BOHLEN UND HALBACH

Photo from European

SMELTING FURNACE OF THE NEW KRUPP STAMPING MILL IN
ESSEN-BORBECK, SHOWING BLOWER, DUST SEPARATOR, AND GAS PIPES.

works in the Ruhr district." This statement hardly bears the stamp of accuracy; the name of the firm would not have disappeared; the family interests in the firm of Krupp were no greater than those of Thyssen in his own concern, neither had Krupp any special character in development and expansion which the others did not have.

The real reason was not disclosed. The Essen firm did not wish to merge their identity in the great combine, because they were still hoping for the day when they would again be given a real "Krupp" task. Would that day come?

SECRET ARMS

Krupp did not propose to let themselves be caught unprepared. In spite of all advertised satisfaction over technical achievements proclaiming that "Krupp are just as great in peace-time work" they still yearned for those other tasks mentioned only in whispers in Essen, for the past decade.

The hope had not been given up, nor had the work. The vigilance of the I.A.C.C. prevented any open attempt to violate the disarmament clauses of the Peace Treaty on the part of the Essen establishments. All that could be done was to keep the plant ready for a quick change to arms production. The rolling mills, steel forging presses, drilling machines, and lathes in use, differed from those destroyed in 1919 only in being more modern and including technical modifications demanded by improvements in manufacturing processes. Identical plants might be seen in Le Creusot and Pilsen. Whole sections of the Essen works were Sleeping Beauties now, waiting the word that would restore them to old time splendour and glory.

Preparations were made with characteristic Krupp thoroughness. The practical was not permitted to outstrip the theoretic-technical. This last was the task of the mysterious "Research Department for Arms Production" which continued to exist as if there never had been a November, 1918, or a Versailles Treaty. Confidential messengers of the firm shook their heads dubiously as, day by day, they carried wagonloads of technical journals and books into the secluded offices of the department. The material was taken in charge by an equally serious staff of technicians and conscientiously studied. Here, at work, were gun constructors, gun-carriage designers and armour-plate specialists; here were created, . . . on paper . . . the newest designs for engines of destruction. Here worked earnest students from families devoted to the firm, learning to make practical use of the experience of their elders, all of them looking forward to a future that still lay in darkness. But even then, the firm had real tasks for them, . . . in secret arms factories outside Germany.

There was first of all little Holland, most conveniently located at the very gates of the industrial district, and for years past the whole of Krupp's overseas exports had been shipped from Dutch ports.

Krupp had secured firm foothold here during the war with its many murky transactions here, there, and everywhere. Since 1916 the Black List of the Allied Powers had included the Hague firm of Blessing & Co., founded by Krupp middlemen, and a firm which became really prominent only after the collapse of Germany, when they figured as large "buyers" of vast quantities of German arms saved from destruction in Essen, Magdeburg, and Düsseldorf. Large depots in Hogezand (Groningen) and Krimpen on the Yssel served to receive ordnance material sufficient for the needs of a whole army, including up to 1500 guns. A change to the name of "Siderius A.-G." of the "Hollandsche Industrie en Handel Maatschappij" was intended to conceal the connection with Krupp. But in 1926 diplomatic representations on the firm's activities were made by France to the Dutch Government who declined to intervene in what they termed "private business" in which influential Dutch nationals were interested. The Siderius A.-G. gradually became a holding company for a number of armament undertakings; they owned the Rotterdam engineering works and shipbuilding yard of Messrs. Piet Smit, Junior, which included a special department for ordnance manufacture, staffed by the constructional department of the Germania shipyard of Kiel, then bodily transferred to Rotterdam. Other subsidiaries were the "Maschinen en Apparaten Fabrik Utrecht" and the "Ingenieurskantoor voor Schepsbouw" of The Hague, both of which produced torpedo parts for the Krupp shipping firm in Rotterdam.

The supreme head of the secret factories in Holland was the Krupp-Director Buschfeld, who made his sixtieth birthday an occasion for an inspired article in the Essen press concerning "his deliberate avoidance of publicity." This reserve in a director of the Amsterdam Crediet Maatschappij and of the Rotterdam "Kruwal" is not to be wondered at. He did not appear on the Board of the Siderius Company either; here the officiating Krupp directors were Siegfried Fronknecht and Henri George, while forty other German engineers were employed there in the technical department.

Special mention should be made of Herr van Beumingen, Chairman of the Siderius Board of Managers, a Hollander used as stage setting for the Krupp interests. Trader in arms in a big way, a dangerous intriguer in murky political business, he was of the Zaharoff-Gontard-Mandl type, the sort of man little known to the general public. As official head of the Dutch selling office of the Ruhr Coal Syndicate, Beumingen had a hand in all the doubtful dealings that rise like poison gas in the swamp of the Krupp secret production. He stood back of the Frank Heine Case, the "Utrecht Forgery" through which a secret French-Belgian pact was negotiated to thrust the sword of deadly suspicion between Brussels and the Hague.

And . . . It can be told now, it was he who, somewhat later, financed the Dutch Black Shirts whose slogan was "Down with Corruption!"

Holland was particularly useful not only as the place of secret manufacturers, but also as a place of consignment for imports from other countries which could not be declared as for Essen. Most of these imports came from Sweden, which supplied the bulk of the iron ore required by the heavy industries of Germany after the loss of the Lorraine iron-ore deposits. In this case Krupp had succeeded in pulling off a brilliant *coup* after the end of the war—they had penetrated into the Bofors Arms Factory. When the Allies stipulated that Krupp should destroy machinery and equipment, they allowed equally dangerous assets in the shape of patents, licences, and secret processes of manufacture to remain in Krupp's hands. Krupp proceeded to sell the most valuable of these assets to the Swedish firm, who paid with a substantial block of their own shares. These were increased by private purchases on the stock exchange with the aid of a substantial grant from the German Government, ostensibly for the purpose of refinancing the Rheinmetall Company. By the end of 1925, out of a total of nineteen million Bofors shares, over six million were in the hands of Krupp, giving a controlling interest in the company. The situation is not devoid of humour. While the Inter-Allied Commission of Control was vigorously and conscientiously "supervising" German disarmament, Krupp was again busy in the Bofors works with the manufacture of the latest patterns of heavy guns, tanks armed with one machine gun capable of firing 1000 rounds a minute, anti-aircraft guns, gas bombs, and many other things.

But Sweden was not Holland. The Socialist party of this democratic country soon discovered who was master of Bofors, and in 1929 the Stockholm Riksdag adopted a law prohibiting foreigners from owning armament works in Sweden. As nothing short of direct action against the actual property of armament interests could establish any semblance of control over their activities, it was obvious that the above legislation would prove futile. Krupp's money enabled them to buy all they required—not excepting reliable dummies. They accordingly proceeded to create a holding company of Bofors shareholders in which Krupp no longer appeared.

The Chairman of the Company, Director Mauritz Carlson, owned 63,000 Bofors shares. But Carlson is heavily in debt, in fact to the amount of 6,300,000 crowns, and . . . so surprising! it is Krupp to whom he has pawned his shares for this amount. The Assistant Manager, D. Wingquist, is in the same position; his large ownership of Bofors shares is held in the Skandinavia Bank for his creditors. Now that all this was arranged so nicely, it was quite easy to swear, in court if need be, that Krupp had nothing to do with the Bofors Company.

How was it possible, though, to get round the disarmament clauses of the Treaty of Versailles in such cynical fashion? If the governments concerned did not want to hear anything about it, why were the great competitors of Krupp, in Le Creusot and in Sheffield, silent? They must have been well aware of Krupp's activities. The answer to these questions is contained in a speech made by the Deputy Paul Faure in the French Chamber on the post-war activities of armament firms: "If we examine this matter closely, we will find that they (Krupp and Schneider) are working together in Poland, just as they did before the War at the Putiloff works in Russia." And another reason may be found in the big advertisements kept up for years in German military and technical papers by the House of Vickers, the true character of which could hardly be hidden. Germany's military spirit was needed, as was Krupp's "secret" arms business; both were needed to wake up the depressed trade in armaments. Where would one be, without the excitement of an occasional secret dossier re Essen's foreign connections?

Had the days of industrial rivalry in armament production gone? With all these pacifist parliaments, disarmament conferences, and outlawry of war, the worthy business men of the bloody trade were careful to avoid doing anything to injure their common interests. It was not always easy for them as, for example, in a case like the dispute between Vickers and Krupp over certain royalties dating back to the days of the War. For four years the British firm daily fitted some tens of thousands of shells with excellent Krupp patent fuses for use on the Allied fronts. Although it cannot be determined exactly how many German soldiers fell victims to their effective explosive working, the extent of the claims made by Krupp for payment for the use of their fuses is known.

Known, and unassailable. At 1 shilling 3d. per grenade Vickers-Armstrong's four-year production summed up to a total of 123 million shillings owing to Essen.

That terrible piece of lying wartime propaganda, that Germany made fats from corpses of soldiers, hereby acquires something very like a sinister reality; the House of Krupp reckoned sixty marks per man for those two million dead German soldiers in computing the account presented to its competitors. The heroes of Langemarck and Ypres rose from their graves—in the shape of a debit of 1s. 3d. in the books of Vickers-Armstrong, Ltd., a settlement of which was demanded by Essen. This business of selling to both sides of a front, hitherto straight earnings from contracts and frequently excused by a lack of political foresight, now acquired a new and more sinister meaning; it amounted to a permanent interest in every shot fired by the enemy and in every casualty suffered by one's own side. Krupp's claim was, however, surpassed by that of their allied firm, the Stumm concern, which in 1913 claimed compensation for being forbidden to mine iron ore in the great military cemeteries of Metz and St. Privat. The dead did not hinder the earning of profits; they actually constituted these profits at the rate of 1s. 3d. for every shell which slaughtered them.

The clash with Vickers involved a certain amount of friction, as £6,150,000 was no trifle, even for Sir Basil Zaharoff's firm. As direct negotiations proved unsuccessful, Krupp appealed to the Anglo-German Arbitration Committee assembled at the request of the Berlin Foreign Office.

That they themselves might appear as witnesses in their own case, Essen ceded the claim to the Industrial Joint Stock Bank, one more dirty touch in the whole unclean business, for the President of the Bank was none other than Dr. Gustav Krupp. No information regarding the proceedings was allowed to be known (the whole affair was not revealed until much later) as both parties were anxious for a discreet settlement.

The Paris journal "Le Crapouillet" says of this settlement: "Krupp did not follow up his claim and was satisfied with the compensation offered by Vickers, *i.e.* an interest in the Steel Rolling Works Miers in Spain."

In Spain! The gentleman of the House of Vickers could feel certain of the acceptance of this offer, as any further growth of Essen's Spanish influence was most important.

The Krupp interests in Spain already included the Maquinista Terrestre y Maritima, the great engineering works of Barcelona, the Levante Shipping Union, and, more recently, their former subsidiary the Orconera Iron Company of Bilbao. The significance of these interests in iron ore, engineering, and shipping companies is explained by the Krupp connection with the Eschevarita shipyard in Cadiz, where a number of Krupp engineers from the Rotterdam office were employed on the construction of submarines which could not be built in Kiel, and in which the improvements made as the result of experience gained with the huge U-boats used in Germany's four years of submarine warfare were embodied.

Only a fraction of these arms and warlike stores made in Holland, Sweden, and Spain found their way to Germany, which was not yet ready for them. To meet this difficulty, well-concealed depots were established in neutral countries, the contents of which were available for immediate use. The lucrative arms trade was also resumed, even if in somewhat dubious ways at the outset. The risings of the Riff tribesmen in Morocco, the Druses in Syria, South American revolutions, and civil wars in China gave rise to persistent rumours of Krupp agents being implicated.

But it is now time to revert to events in Germany leading to developments which were to afford Krupp the best oppportunity they had had since the World War of playing a leading part in the destinies of the country!

THE CRISIS YOU NEED

Towards the middle of May, 1929, a luncheon party was held at Hügel, to which a number of prominent industrialists were invited by the host at the suggestion of the President of the Reichsbank, Dr. Schacht. The latter had just come from Paris where he and Albert Vögler, Managing Director of the Vereinigte Stahlwerke, had been attending a meeting of experts to discuss a revision of the Dawes Plan and at which American creditors were represented by Messrs. Owen D. Young and Pierpont Morgan. Germany was facing serious difficulties over the question of reparations, as beginning with the next year she would have to pay the full annual rate of 1250 millions and that at a time when the boom was beginning to wane and the first signs of the coming slump were discernible.

The Ruhr industrialists, old Kirdorf, Reusch of the Good Hope Ironworks, Klönne, Hans Cremer, and Fritz Thyssen, besides their host Krupp, carefully listened to what Schacht and Vögler had to tell them concerning fresh demands made by the creditors; payments to be continued until 1988 at the rate of 2000 millions per annum, of which amount 600 millions were to be paid unconditionally and the balance subject to transfer protection.

Dr. Schacht strongly urged the acceptance of these proposals, later to be known as the "Young Plan," although actually initiated by the Frenchman Quesnay. Their advantage was the provision of the transfer clauses, as the continued payments at the rates laid down by the Dawes Plan would inevitably lead to a financial catastrophe, involving the collapse of a great part of German economic life.

The words "catastrophe" and "collapse" made Fritz Thyssen sit up. This eternal son of a great father watched his inheritance dwindling alarmingly. The crisis brought about in the Steel Trust with its liabilities of 700 millions, red tape over-organization, and defective rationalization was not yet outwardly visible, but insiders knew well enough that the true balance sheets would not bear examination. Thyssen thought that if he must go bankrupt, he might as well contrive to make others responsible for it:—

"This is the crisis I need just now! It is the only chance of disposing of the wage and reparations questions once and for all."

Disposing of the wage question? The Government was still that of Hermann Müller, with Hilferding as Minister of Finance. The Socialists were still the strongest party in a parliament which had just moved further to the Left. But here, at this luncheon party in the castle of Hügel, a cynical recipe for "disposing" of the burden imposed by these unpleasant realities by means of the crisis "which I need," by bankruptcy, agitation against reparations, and political difficulties on a scale exceeding those caused by the Ruhr campaign, was coolly put forward.

Idyllic dreams of a bankrupt? Tall talk? All that and more. In such company and in such a place the call for a crisis became a programme. Was it a reflection of the first dull opposition aroused throughout the country by disappointment for a prolonged period and bound up with the whole problem of Reparations and "Systems"? Both were coupled together that one might dispose of the other . . . that was the general outline of policy sketched out at that fateful luncheon party as guide for some of them at least. Vögler resigned some days later; that same autumn Kirdorf watched the march of the National-Socialist Storm Detachments in Nuremberg; Thyssen invited Göring to a preliminary talk in his castle of Landsberg; and finally Schacht, having learned much at Hügel, commenced his violent attack on the Young Plan and the Young System. The gentlemen had understood each other perfectly. When Georg

Bernhard revealed the gist of Thyssen's remarks in the Reichstag, there was no lack of persons to refute this dangerous statement.

Was Krupp with them? His attitude was not quite clear. Ties with existing Government circles were closer than those of his colleagues, even if affected by connection with the Ministry of War. And the final incentive failed; Krupp was not bankrupt. During the past boom years no dividends had been paid—a measure of shrewd foresight which involved no sacrifice for a family concern. Although the head of the firm had usually displayed more caution than foresight, he had acquired a reputation as a prominent leader of industry—a pitiful illustration of the barrenness of the epoch now closing!

But even Krupp realized that they were on the eve of far-reaching changes. During the great labour crisis in the metal industries, which occurred in 1928, *i.e.*, twelve months earlier, Krupp forsook their habitual attitude of reserve and openly joined the oppressors. The Government of the Social Democrat Hermann Müller had dared to give State assistance to the 100,000 locked-out workers; for this it was never forgiven. And it was this fact which drew Krupp, Haniel, and Siemens, the bankrupt Steel Trust, and the coal-mining companies together. The "political remuneration" the Socialists had succeeded in assuring for the workers greatly alarmed all industrialists. If political power given to workers was to be used by them as a lever for safeguarding and improving their economic position, it would become all the more necessary to offer strong opposition to Socialist influences in the Government and to guard against the possibility of the Socialists attaining power in the future. Although most of those concerned were certainly reluctant to face the coming crisis in the same spirit of optimism displayed by Thyssen, the determination to dispose of the wage question by political means was common to all.

When in the autumn of 1931 Dr. Duisberg of the I. G. Farben was succeeded by another man from the Ruhr, Krupp von Bohlen, as President of the Federation of German Industries, it may be said that the prolonged period of neutrality in Germany's industry had come to an end.

The selection of Krupp indicated a compromise. While the bankrupt firms had already fled to cover by donning Brown Shirts, the greater number of German industrialists were only ready to go half-way in the great political upheaval arising out of the Hitler victory at the polls in September, 1930, dashing wave after wave against the foundations of the German Republic. The disclosure in the Reichstag, by the member Breitscheid, of a resolution adopted by the Ruhr mining industry to levy a sum of fifty pfennigs per ton for political purposes, and to assign 600,000 marks from the fund thus realized to the National Socialists, was a grave symptom, but nothing more.

In Harzburg, where Hugenberg and Hitler made their first public appearance together, only second-grade directors and legal representatives of the great industrial families were present, while the principals themselves were conspicuous by their absence. Again at the Industrialists' Conference in Düsseldorf, where the head of the Nazi Party first explained his programme, the "Heil, Herr Hitler" uttered by Fritz Thyssen produced little response from the remainder of the assembly. And the new Federation President, Krupp, in declaring that Germany's industry was overburdened by the cost of its own inner organization, made what was very like an indictment of Thyssen's earlier talk at Columbia University, New York, about the German system of taxation. In what direction were Krupp and the majority of German industrialists moving in the autumn of 1931? Towards the moderate course steered by Brüning or towards the dictator plans emanating from certain sections of the Right or from the Chancellery of Hindenburg? One thing is quite certain: they were not yet moving towards Hitler.

The stage was, however, being set for events which were to change the position overnight and which would compel these reserved industrial magnates to make a move. At the end of 1931 the Brüning Government received confidential information that the Crédit Lyonnais, the great French industrial bank, was negotiating with the hard-pressed shareholder of the Steel Trust, Friedrich Flick, in regard to the purchase of that block of Gelsenberg shares which, in the singularly involved circumstances prevailing amongst the Ruhr undertakings, may be said to have constituted the key position in the Steel Trust. This information did not come from Düsseldorf or Essen, but, quite casually, from Paris itself, and was rendered all the more significant by the supplementary intelligence that the bank was acting on the direct instructions of Schneider-Creusot. The French armament concerns were anxious to acquire control of the largest German steel undertaking.

The horrified Government promptly swallowed the bait and hastened to prevent the worst. Finance Minister Dietrich entered into negotiation with Flick who, now that he was sought after, became less amenable. He finally agreed to cede his holding of Gelsenkirchen shares at a price of ninety marks per share, *i.e.*, at over four times their market price, which was actually quoted at about twenty. On March 4th, 1932, the deal went through at a total figure of 110 millions for the entire block of shares.

The State was now the owner of the Steel Trust, the greatest and most central complex of heavy industry, controlling no less than seventy-five per cent. of Germany's total production of iron ore and fifty per cent. of her output of coal!

The spectre of nationalization arose before the eyes of German industrialists. Who could tell if the Government would be content to play the part of a non-active shareholder? Since that sensational case with the Hybernia Mining Company, taken over by an imperial minister nearly thirty years earlier, the resentment felt by the Ruhr magnates against the Berlin bureaucracy had never quite died down. In the face of the anti-capitalist feeling generally prevailing throughout Germany, even a simple manœuvre to secure a Government grant, as desired by the principals of the Gelsenberg Company, might readily be turned into a dangerous precedent. The control now exercised by President Hindenburg's Chancellery was not strict enough to prevent any such action on the part of the Government. To prevent a return to a Government of the Left, which would then be placed in a position of unassailable strength, it would be necessary to act more vigorously than heretofore. The acquisition of the Vereinigte Stahlwerke sounded the death knell of parliamentary democracy; even the hitherto neutral industrialists were now hostile. The sales agreements had only just been signed when an energetic swing to the Right swept away the Brüning Government.

Papen's political programme, introducing his Cabinet, was aimed at State Socialism and was intended to placate alarmed industrialists who understood it well. Among the many contacts possessed by this clever gambler during the great crisis, those with the castle of Hügel were kept curiously dark. The closeness and extent of these ties were revealed by the circumstances which brought about the fall of the Prussian Braun-Severing Government in July, 1932, when even the relatively moderate Krupp approved and supported the attack made on the remnants of German democracy. Was it wholly fortuitous that the leading actors in this violation of the Constitution were all persons closely connected with Krupp? The new Commissioner for Prussia, Bracht, formerly Burgomaster of Essen, had been a confidential adviser of the Krupp family for the past ten years; he was a zealous upholder of Krupp interests and was also an executor of the late Hugo Stinnes, and the guardian of his children. Bracht's satellites were drawn from the same source, the Chief of Police of Essen, Melcher, and General Rundstedt of the Westphalian Army Command, both being dyed-in-the-wool reactionaries always courted by industry. On the eve of the *coup d'état,* Bracht came from Berlin to Essen for a few hours to confer with the Master of Hügel.

The ties which bound Essen to General Schleicher were just as close. In this case it was Otto Wolff, a partner in many post-War undertakings of Krupp, who was responsible for them. The all-powerful chief of the Bendlerstrasse meant a good deal more to Krupp than a mere hope for the future, as the Third Reich's drive

Photo from European

DR KRUPP VON BOHLEN UND HALBACH, LEADER OF THE GERMAN INDUSTRIAL ASSOCIATION, ADDRESSED A MEETING OF 20,000 REPRESENTATIVES OF GERMAN INDUSTRY SUPPORTING THE HITLER GOVERNMENT.

Photo from European

COOLING TOWERS FOR THE BOILERS IN THE KRUPP PLANT AT ESSEN.

toward rearmament dated from Schleicher's coming to power. In the summer of 1932, when a demoralized Germany was facing the future with complete apathy, the first refreshing breeze penetrated into the Krupp works and turned their dreams of rearmament into something approaching reality. The beginning of preparations to increase the Reichswehr to 300,000 men brought important armament contracts to Essen for the first time in many years. They included orders for ordnance material and skilfully disguised component parts of tanks, not to mention gas masks to the value of a million marks—the first swallow of the new summer of armaments!

And Hitler? Were there no actual connections between the Essen firm and the man who was already on the threshold of power and was shortly destined to mean so much to the future of Krupp? The answer to this question cannot be based on conclusions drawn from appearances. Although there were many reasons why they should have been drawn together, yet they were not. This circumstance was due to a variety of complicated relations and contradictory causes which were not always readily discernible.

The seeds in the ground which produced and nourished Hitler were sown by Krupp. The German Workers' Union, in which Hitler appeared in 1919 as a spy of the Reichswehr and which he subsequently transformed into the National-Socialist movement, was a Munich continuation of the well-known "Vaterlandspartei" founded towards the end of the War with the financial support of Krupp. This connection is not merely on the surface. Nearly twenty years of unscrupulous propaganda, financed by those interested in convincing the *bourgeois* section of the German people that their country's salvation was dependent on a policy of naval and military armaments and of territorial annexations, had found a worthy successor in the person of the man from Braunau. He was what Schweinburg and Keim were, but in an altered world.

Were there any financial connections between Krupp and the Munich group in the early days of the Hitler movement? Surprising as it may seem, it is a fact that Krupp money was particularly plentiful in the Munich of the early post-War years. In the course of the trial of the assassin Lindner in December, 1919, Chief of Police Staimer stated that he was once informed by a person possessing his entire confidence that he had been offered 50,000 marks "for combating Bolshevism" with the assurance that no accounts regarding the expenditure of the money need be rendered; notwithstanding his refusal of the offer, this person declared that he had actually had 10,000 marks of this money in his hands, and the paper wrappers round the banknotes bore the impress of the firm of Krupp. Later, banknotes in similarly stamped wrappers were found in the possession of a leader of a reactionary group of plotters, named Lotter.

It was in this political underworld of the earliest organized movements by counter-revolutionaries, where Krupp money circulated freely, that Hitler's political career had its beginnings. With the assistance of his supporters in the Reichswehr, money from the shadiest sources flowed into his pockets, his chief patrons being Bavarian industrialists, chief among them the Munich branch of Krupp. Apart from the reference made to Krupp by Staimer, the name of the Essen firm did not figure in the political chronicle of corruption of the next fourteen years, neither was it to be found amongst those of Hitler's financial backers. No hasty conclusions should, however, be drawn from these facts; the "silent force" of Krupp, as it was once called, had not ceased to exist. The Essen firm, for reasons which will be referred to later, displayed considerably more prudence in advertising its activities than its Big Industry party friends. This applied to Hitler just as much as to the remaining organizations of the Right, such as the Stahlhelm (Steel Helmet) and the Vereinigten Vaterländischen Verbände (United Fatherland Associations), together with the Black Reichswehr. The favourite form of financial support given from Essen through roundabout channels and intermediaries was lavish subscriptions on behalf of one or other of the industrialists' associations. No important subvention for Hitler from the Ruhr industrialists could be made without the connivance of Krupp. Its representatives on the Executive Boards of the various federations held such power that any money given Hitler by any of these organizations, even in the years of political differings, must be booked as an evidence of good will on the part of the Essen House.

Hitler certainly regarded it as such. Although in the course of his address to the Nazi party conference held in Nüremberg in 1927 he complained bitterly about the obtuseness displayed by the leaders of German industry, and on other occasions uttered warnings about the desirability of their changing their attitude, yet he invariably referred to the aloof House of Krupp in terms of courtesy. In the official commentary on the National-Socialist party programme, "the really great pioneers of our heavy industry, Krupp, Kirdorf, Thyssen, Abbé, Mannesmann, and Siemens" were reverently excepted from the proposed nationalization of industry. On the occasion of his last conversation with Otto Strasser, Hitler replied to his question: "Would you, for instance, leave everything just as it is in the case of the Krupp A.-G.?" with an unmistakable: "Why of course!"

The reason why, notwithstanding this mutual exchange of esteem, no direct contact was established between Krupp and Hitler (who was doing his utmost to realize the Cannon King's dreams) before 1933, was the peculiar position of the Essen firm, their administration, and their traditions. Consideration for their relations to the democratic Government may have also counted as inhibition. But, judging by precedents in the history of the firm, the bridging of the gap which separated the conservative-monarchist connections of an old-established firm from the demagogic ruthlessness of a social outcast thirsting for power, did not present insuperable difficulties. Most important, however, was the attitude of the Reichswehr, to which Krupp were bound by countless ties of official and unofficial co-operation, and as long as the leading men in the Bendlerstrasse in Berlin were not unreservedly backing, or at least recognizing, Hitler, the Essen firm would also keep their distance. These fourteen years of republican and democratic rule constituted a diplomatic period in which Krupp had to make an outward appearance of disavowing a movement with which they were in complete sympathy of aims if not methods.

Down to the very hour at which Hitler seized power in circumstances which have not yet been fully explained, Krupp continued to back the last great opponent of the Third Reich, General Schleicher. An attempt to invite Krupp von Bohlen to Berlin in order to bring pressure to bear on the old gentleman in the Presidential

Palace came too late to prevent the lie about the mobilization of the Potsdam Reichswehr. Krupp was barely in time to declare his adherence to the victors of January 30th, 1933, thereby displaying less acumen than his confidant Bracht, although his friends Papen and Hugenberg were in the party.

When General von Blomberg contrived to bring about the great reconciliation between Hitler and the Reichswehr, the Swastika flag was hoisted even over the castle of Hügel.

The "counter-revolution" the shrewd old Cannon King had foreseen, had come to pass—if in a somewhat different form from that which the majority of German industrialists expected it to take.

FORWARD OVER GRAVES

It was good to be alive in that spring of fulfilled hopes of 1933, the year one of the rebirth of Krupp.

Already in February the revived General Staff resurrected secret file "K" in order to approve this work of many years of leisure. At No. 9 Margarethenstrasse, Berlin, the Central Bureau for German Rearmament had been established, and was sending urgent summonses to the arms works throughout the country.

It was a race between Germany's new armaments and the awakening of neighbouring States to realize the situation. Germany was passing through the "danger area" of her programme of rearmament.

The pace was terrific.

Feverish activity prevailed on the reloading wharfs of Antwerp and Rotterdam, where cargoes of raw materials destined for Germany's new armaments in the first four months of 1933 vastly exceeded the entire tonnage of the previous year and soon amounted to staggering totals. The imports of copper went up from 8000 to 15,000 tons, those of scrap iron from 10,000 to 83,000 tons, and those of iron ore from 35,000 to 208,000 tons, including, for the first time since 1914, zircon ore from Brazil of a grade used only in the manufacture of gun steel.

At the main Krupp works in Essen 5000 new specialist workers were taken on within a few weeks, mostly skilled foundrymen, turners, and draughtsmen drawn from the ranks of skilled labour of the days of the World War. At the rolling mills the peace-time plates gave way to heavy armour plate; thousands of precast blocks of steel were transformed into gun tubes; shell cases appeared in the lathes, and the trucks of yesterday were replaced by grim-looking tanks.

On the range at Meppen (now it was worth while having kept it fourteen useless years) huge howitzers were being tested; strong enough to batter the concrete ring behind the Vosges to pieces. In Jüterbog quick-firing howitzers with motorized traction were being calibrated, and six-ton tanks, the first of a series put into production at Magdeburg, manœuvred over the fields. Anti-aircraft guns were being tested on the Baltic Coast.

The Germania shipyard screened their long deserted slipways from prying eyes and began the construction of submarines, minesweepers, and destroyers launched at the rate of one a month. Five great destroyers, each 312 feet long, were put in hand. In the machine shops of the yard overtime work hastened the completion of 600 heavy motors for marine tanks, steel amphibians capable of traversing any river or sea inlet.

All technical resources of this busy anthill of the Krupp concern which were not otherwise engaged were used for high-speed manufacture of tens of thousands of trucks and tractors for the mechanized units of the Army. It became known that the Vice-Chairman of the newly formed "Reichsautobahn Gesellschaft" (National Motor Road Company), hurriedly established for the express purpose of dealing with the problem of military road transport, was Baron von Wilmowski, brother-in-law of Dr. Krupp.

The beginning was promising. By the end of 1933 the number of employees had risen by 11,500, the greatest single jump in the labour establishment of the Krupp Company. The turnover increased by about fifty millions—with reduced rates of pay. The men's trade unions were destroyed and their leaders were dead or in concentration camps, despite the fact that some of them were much sought after and courted in November, 1918, and that these very men not infrequently represented Krupp interests in negotiations with the Government. The Krupp management hailed the resultant powerlessness of their workers as a state of genuine social harmony "according to old Krupp tradition." In the spring of 1934 Hitler appointed the lord of Hügel as "Leader of Economics," and placed the first seven groups of the new economic organization corresponding to the entire industry of Germany under his control.

Had Krupp become reconciled to the Third Reich? If immediate advantages are to be weighed against wider interests, he had every reason for so being. The works were overwhelmed with contracts; the semi-legal character of armament production was already a polite fiction about to disappear completely; profits were being made, and Krupp was once more master in his own house. What more could he want?

Krupp's real wishes are revealed from the general tenor of his public speeches during this first year of the Hitler régime. On his appointment as Economic Leader, he announced: "Our aim is to effect a co-ordination of the Totalitarian State and the responsible independence of industrial administration." Co-ordination? No unconditional subordination therefore and no mention whatever of nationalization. On another occasion Krupp said: "An industry without profits is unable to provide permanent employment." This was aimed at the jugglers with unemployment returns, whose paper schemes dictated increases in employment; and at the economic nostrums of the various quacks who had acquired momentary prominence through the national political upheaval. Krupp struck a more serious note in a speech delivered at an anniversary celebration: "However good a national industrial economy may be, it cannot dispense with international connections and trade. When we sell German products of high quality to foreign countries, we are furthering national interests." Who had ventured to cast such aspersions on the "national" attitude of the House of Krupp that its Head found so vehement a defence necessary? Perhaps his reply was intended for the preachers of autarchy, or those sharp critics still to be found in the ranks of the Nazi party, who complained about a very definite class of "exports to foreign countries" by Krupp—armour-plate licenses, and special steel processes, the newest and most valuable: in 1931, to the United States Steel Corporation, and in 1932 to the English firm Dorman Long & Co., Ltd.

Thus far the reproaches and retorts had been veiled. But there was no doubt about the nature of the pronouncement made by a certain powerful and influential new director of Krupp, the elegant and ever-smiling Professor Dr. Görens, to no less a person than the great Göbbels himself.

"Before building, make certain that the ground is firm enough for your foundations. The State also needs firm and reliable ground to bear its foundations with safety. In our German Fatherland this ground is represented by a great tradition."

What was this foundation to which the Krupp Director referred? What was it that must be "firm" and "reliable" in order to secure the "safety" of the State? Was it public order or subordination?

We come to June 30th, 1934.

First, an eposide which took place three weeks earlier. On June 4th, at the instance of Chief of Staff Röhm, a delegation of S.A. men arrived in Essen for the purpose of visiting the Krupp works. Although a notice at the main entrance clearly stated: "It is requested, to obviate misunderstandings, that no applications to visit the works be made, as such applications cannot, under any circumstances, be granted," it was obvious that visitors of this class could not be denied admission. Professor Görens received the delegation, which was headed by the Chief of the Political Bureau of Storm Detachment, Supreme Headquarters, von Detten. The visitors were shown over the works and no objections were raised when they suddenly expressed a wish to address an assembly of the S.A. men employed by Krupp. The speech made by von Detten was a message; he stated he had come to tell them that, although Chief of Staff Röhm was a long way off from these labouring comrades, yet he never lost sight of their interests and welfare. The smile vanished from the face of Professor Gorens. The warning was not lost on him. The delegation were obviously beating up a following for a second revolution.

On June 29th Hitler himself arrived in Essen, ostensibly in response to an invitation from Göring to attend the wedding of their mutual friend Terboven. Ever since March, 1933, Goring had made the place a private base for his own power, and he hoped to spur the hesitating and irresolute Hitler into action more effectively in Essen than elsewhere. Sensational police reports submitted to Hitler contained details of dangerous schemes and plots of a secret camarilla of Röhm, Schleicher, and Strasser, and it was necessary that a demand for vigorous action on the part of the still undecided Hitler be pressed with all speed.

Dr. Gustav Krupp von Bohlen was present at the conference.

In the entrance hall of the Works Fräulein Irmgard von Bohlen presented a bouquet to the Führer. The assembled directors and engineers of the firm raised their right hands in the German salute, but if Hitler looked at their faces he saw little devotion and less enthusiasm reflected in their expressions. The "Kruppianer," as they call themselves, are more moderate, sober, and critical than people in other industrial centres.

Hitler let himself be conducted through the huge Works. He was particularly interested in the departments concerned with armament production. He was accompanied by the head of the firm who, the foreign press announced some days before, was desirous of resigning his post of Economic Leader. It was, in fact, merely a warning to Berlin and the report was denied in Essen as soon as its effect had been achieved. Hitler now knew that the man whom he was then following through his industrial stronghold was dissatisfied and suspicious.

Not for the first time were far-reaching political decisions taken in the private room of the head of the Krupp works. Bismarck before going to war, and William II in the summer of 1918, both sought the atmosphere of Krupp steel at a critical time. On this 29th of June, 1934, serious news came from Berlin: the Minister of National Economics, Schmidt, had gone on leave, a thinly disguised resignation; the meetings of the Chamber of Industry and the Chamber of Commerce received a bombshell in the shape of the sharp transfer note from the United States; the danger of a complete collapse of industry bolstered up by artificially created employment of labour alarmed all industrialists. This general unrest! His Excellency Herr Krupp reminded his visitor of the dangerous symptoms of a "second revolution" foretold by highly placed Government officials.

They now proceeded to discuss the main point. What was said on both sides will long remain hidden in the private archives of the Essen firm. But undoubtedly Krupp then obtained confirmation of the frequent assurances made to him by Fritz Thyssen in the Industrialists' Club (the "Ruhrlade"); even this *arriviste* hated revolutions like poison. He made the declaration on taking office and he has invariably adhered to his attitude during all the years of his rule. In the present case he was not breaking faith—on the contrary, he was keeping faith when he assured his companion of his determination to protect German industrial enterprise against all attacks and of his inflexible resolution not to allow a second revolution, and so forth.

Did Hitler confide any part of his plans for the next day to the head of the Krupp concern, President of the Federation of German Industries and Leader of Economics?

Time sequence is always an important index in criminology. And here it points compellingly to knowledge and responsibility of those circles whose voice for the moment was the cool correct man in the head office of the Essen House. When Hitler entered the plane for Munich in the night from June 29th to 30th, he took with him certainty that the Western industrial magnates would stand by him in his murderous blow against Röhm. This time there can be no doubt as to who led the Leader.

Forward over graves! The blood bath of this German St. Bartholomew's Night which sealed the triumph of the generals over the men in the ranks, of the industrial magnates over Socialist visionaries, and of business over chaos, ushered in a new era. Normal standards of patriotism had been re-established, and from now on every political decision could be recorded on the credit side of the Krupp ledgers; the proclamation of open rearmament, the fortification of the Rhine and the Anglo-German Naval Agreement. They are marked by the milestones of Krupp progress, in terms of the gross profits shown in their balance sheets—

1932 - - - -	108 million marks
1933 - - - -	118 million marks
1934 - - - -	177 million marks
1935 - - - -	232 million marks

Does all this National Socialism make us dizzy? These millions do not tell the whole story. The scanty information contained in the involved published accounts of the firm of Krupp reveals an actuality which put all the past history of this house in the shade. In 1935 their expenditure on new plant was forty million marks, a greater expenditure than in any one of the four years of the War. Extensive new buildings and plant were provided in Essen and Rheinhausen, and there was also a systematic transfer of certain important works into the interior of Germany where there would be less danger from hostile bombers. The vast accumulation of supplies of essential raw materials ran to the value of eighty million marks, with huge stocks of iron ore, zinc, copper, and nickel, just as in 1914. Large-scale transfers of capital took place, with reserves totalling nearly 100 million marks, or more than half the total share capital. A body of employees increased from 40,000 to 90,000 in barely three years, no longer a restless mass of workers struggling for improvement in their political and social positions, but a militarily drilled army, bound by implicit "obedience" to the "Works leadership."

What was the purpose of that closely guarded department of the Gestapo (Secret State Police) which was now established inside Krupp's Central Office? Why did two workers in Kiel and one in Essen suffer death at the hands of the headsman? The monster gun of the World War had a range of over eighty miles and could bombard Paris from Laon; how far have technical improvements effected by Krupp come since then? The distance from Dresden to Prague is eighty miles, from Flensburg to Copenhagen 145 miles, and from Tréves to Paris 210 miles. It is for experts to decide what is possible there.

For our purpose it will suffice to refer to established facts, rather than to accounts of fiendish inventions described in various secret reports emanating from Essen, barrage guns capable of maintaining a continuous barrage of fire over wide zones of air or land; armoured cars of one and a half, three, and six tons, designed for rapid advances, gas attacks and bombardment; giant submarines to carry such tanks to enemy shores; bombs filled with gas, poison, and bacilli; armour for converting Germany's frontiers into a grey wilderness of steel; giant turrets—but enough of these things!

In the autumn of 1933 the colony of arms dealers in Shanghai, who were buzzing round the festering wounds of China like flies, whispered sensational news—Krupp were offering to supply complete equipments of armaments of all kinds at prices ten per cent. below those of their competitors, with three years' credit and immediate delivery. What did this mean? Essen was having a sale. It was seeking to dispose of stocks of arms manufactured in Spain, Sweden, and Holland and which, up to a few months before, were carefully hoarded reserves, but now were of no value to Germany. She required arms of the very latest pattern, incorporating all the newest technical improvements devised by science for bringing death to mankind.

Was Krupp von Bohlen satisfied at last?

In December, 1934, he resigned his offices in the German industrial organizations, and whenever he has made any public speeches since then he has always stressed the need for increasing export trade. The head of the Essen firm is quite obviously desirous of keeping out of politics. He is no longer satisfied with bayonets as a foundation for increased industrial development. Dwindling exports and, more especially, the loss of the Russian market, were no trifles, even for a busy armament works.

In the spring of 1936 there was news from Jugoslavia; the Belgrade Government was planning to erect iron-works at Zenica and to build a great bridge, costing altogether 200 million dinars, the whole of the contracts to be placed with Krupp. The great European steel companies were caught unawares, but a financial newspaper in Prague announced that Krupp had obtained a promise of the contracts, although their tenders were higher than those of other firms. The London *Times* revealed that Dr. Schacht had deliberately increased Germany's debts in the Balkans and was threatening to devalue the mark, so that the alarmed Jugoslav Government was driven to place the contracts in Germany, to save at least a portion of its money. Blackmail? Merely an honest bit of Krupp business, instigated by an honourable Government.

The spring of 1936 closed a lengthy discussion in the German press concerning the future organization of supplies in wartime. It concluded with hymns of praise for the private armament industry. All Germany was again unanimous in deciding that the private interests of the firm of Krupp coincided ideally with those of national defence. And it was in that same spring of 1936 that Hitler addressed tens of thousands of Krupp workers in the huge Hindenburg Bay of the Essen works. Dr. Gustav Krupp von Bohlen und Halbach sits alongside him, and "publicly honours our great leader Adolf Hitler, to whose service he pledges himself."

BIBLIOGRAPHY

History of the Firm

Wilhelm Berdrow, "Die Familie Krupp in Essen von 1587 bis 1787," Essen 1931 (Berdrow I).

"Friedrich Krupp, der Grunder der Gussstahlfabrik, in Briefen und Urkunden," Essen 1915 (Briefe I).

"Alfred Krupp," Berlin 1927 (Berdrow IIa und b).

"Alfred Krupp's Briefe," Berlin 1928 (Briefe II).

Diedrich Bädeker, "Alfred Krupp und die Entwicklung der Gussstahlfabrik zu Essen," Essen 1889 und 1912.

Fried. Krupp A.-G., "Krupp 1812-1912," Jena 1912 (Jubiläumswerk).

"Die Krupps," Essen 1912 (Jubiläumsheft).

"Erwiderung auf das Rundschreiben der Rheinischen Metallwaren und Maschinenfabrik," Essen 1905.

Wiedfeldt, "Friedrich Krupp als Stadrat in Essen," Essen 1902.

Gustav Köpper, "Das Gussstahlwerk Fried. Krupp und seine Entstehung," Essen 1897.

Hermann Frobenius, "Alfred Krupp, ein Labensbild," Dresden 1889.

F. Wessel, "Krupp," Neurode.

Hermann Hasse, "Krupp in Essen, Die Bedeutung der deutschen Waffenschmiede," Berlin.

Otto Hué, "Krupp und die Arbeiterklasse," Essen 1912.

Wilhelm Düwell, "Wohlfahrts-Plage," Dortmund 1903.

Gaston Raphael, "Krupp et Thyssen," Paris 1925.

T. Kellen, "Friedrich Alfred Krupp und sein Werk," Braunschweig 1904 (Kellen I).

"Die Firma Krupp und ihre soziale Tätigkeit," Hamm 1903 (Kellen II).

I. Meisbach, "Friedrich Alfred Krupp," Köln 1902.
"Der Fall Krupp," Müchen 1903.
H. v. Perbandt, "Ist die Monopolstellung Krupps berechtigt?" Berlin 1909.
Heinz Eisgruber, "So schossen wir nach Paris," Berlin 1934.
H. Murray Robertson, "Krupps and the International Armaments Rings. The Scandal of Modern Civilisation," London 1915.
Joachim v. Kürenberg, "Krupp, Kampf um Stahl," Berlin 1935.

Industrial History

Kurt Wiedenfeld, "Ein Jahrhundert rheinischer Montanindustrie," Bonn 1916 (Wiedenfeld I).
"Das Persönliche im modernen Unternehmertum," Leipzig 1920 (Wiedenfeld II).
Conrad Matschoss, "Ein Jahrhundert deutscher Maschinenbau," Berlin 1922.
H. Thun, "Die Industrie am Niederrhein und ihre Arbeiter," Leipzig 1879.
Josef Winschuh, "Der Verein mit dem langen Namen," Berlin 1932.
Alfred Bädeker, "Jahrbuch fuer den Oberbergamtsbezirk Dortmund," Essen 1926.
Deutscher Metallarbeiterverband, "Konzerne der Metallindustrie," Stuttgart 1924 (D.M.V.I).
"Die deutsche Schwereisenindustrie und ihre Arbeiter," Stuttgart 1925 (D.M.V.II).
Gaston Raphael, "Hugo Stinnes," Berlin 1925.
Richard Lewinsohn (Morus), "Die Umschichtung der europäischen Vermögen," Berlin 1925.
Albert Weyersberg, "Johann Abraham Henckels," Münster 1931.
Karl Mews, "Geschichte der Essener Gewehrindustrie," Essen 1909.

Otto Göpel, "Essen, Montanindustrielle Entwicklung und Aufbau der Ruhr-Emscherstadt," Essen 1925.

Die deutsche Eisen- und Stahlindustrie 1933. "Das Spezialarchiv der deutschen Wirtschaft," Berlin 1933.

Rudolf Martin, "Jahrbuch des Vermögens und Einkommens der Millionäre," Berlin 1912.

Jules Huret, "In Deutschland," 1907.

Armament History

Eckart Kehr, "Schlachtflottenbau und Parteipolitik 1894-1901," Berlin 1930.

"Soziale und finanzielle Grundlagen der Tirpitzschen Flottenpropaganda," in *Gesellschaft*, 1908, 9.

Wolfgang Hallgarten, "Vorkriegsimperialismus," Paris 1935.

"Krupp und Genossen," in *Pariser Tageblatt*, 21. Juli 1934.

"La Signification Politique et Economique de la Mission Liman von Sanders," in *Revue d'Histoire de la Guerre Mondiale*, January 1935.

Otto Lehmann-Russbüldt, "Die blutige Internationale der Rüstungen," Berlin 1933.

Engelbrecht, N. C., and Hanighen, F. C., "Merchants of Death," New York 1934.

George Seldes, "Iron, Blood and Profits," New York and London 1934.

L. Launay et J. Sennac, "Les Relations Internationales des Industries de la Guerre," Paris 1932.

Général Maitrot, "La France et les Républiques Sud-Américaines," Nancy 1920.

Paul Faure, "Les Marchands de Canons contre la Paix," Paris 1932.

W. v. Tirpitz, "Wie hat sich der Staatsbetrieb beim Aufbau der Flotte bewährt?" Leipzig 1909.

A. Saternus, "Die Schwerindustrie in und nach dem Kriege," Berlin 1920.

Alldeutscher Verband, "20 Jahre alldeutscher Arbeit und Kämpfe," Leipzig 1910.
Karl Haussner, "Das Feldgeschütz mit langem Rohrrücklauf," München 1928.
R. Willie, "Ehrhardt-Geschütze," Berlin 1908.
H. Müller, "Die Entwickelung der Feldartillerie von 1815 bis 1870," Berlin 1893.
Henri Bordier, "L'Allemagne aux Tuileries de 1850 à 1870," Paris 1872.
Documents Authentiques Annotés, "Les Papiers Secrets du Second Empire," Bruxelles 1871.
Lewinsohn R. (Morus), "Das Geld in der Politik," Berlin 1930.
A. Nichols, "Neutralität und amerikanische Waffenausfuhr," Berlin 1931.
Rudolf Fuchs, "Die Kriegsgewinne," Zürich 1918.
Heinz Schmid, "Kriegsgewinne und Wirtschaft," Berlin 1935.
Heinrich Ehrhardt, "Hammerschläge," 1922.
Leonidoff, "Die Politik des Rüstungskapitals," in *Kommunistische Internationale*, 1928, 43.

General

"Die Grosse Politik der europaischen Kabinette 1871-1914," 40, Bände, Berlin 1922-1927.
"Die Verhandlungen des deutschen Reichstags," Amtlicher Bericht.
Alfred v. Waldersee, "Denkwürdigkeiten," Stuttgart 1922-23.
E. v. Heyking, "Tagebücher aus vier Weltteilen," Leipzig 1926.
Hans v. Tresckow, "Von Fursten und andern Sterblichen," Berlin 1922.
Fred W. Wile, "Rings um den Kaiser," Berlin 1913.

Hermann Kantorowicz, "Der Geist der englischen Politik," Berlin 1929.

Heinrich Friedjung, "Das Zeitalter des Imperialismus," Berlin 1919.

Ernst Haux, "Was lehrt uns der Krieg?" Essen 1918.

H. Nicholson, "Die Verschwörung der Diplomaten," Frankfurt 1930.

Alexander Conrady, "Die Rheinlande in der Franzosenzeit," Stuttgart 1922.

Rudolf Göcke, "Das Grossherzogtum Berg unter Joachim Murat, Napoleon I und Louis Napoleon," Köln 1877.

Konrad Ribbeck, "Geschichte der Stadt Essen," Essen 1915.

Newspapers and Journals

The Times, London; *Journal Officiel*, Paris; *Berliner Tageblatt*, Berlin; *Vorwarts*, Berlin; *Deutsche Allgemeine Zeitung*, Berlin; *Germania*, Berlin; *Tägliche Rundschau*, Berlin; *Freisinnige Zeitung*, Berlin, *Frankfurter Zeitung*, Frankfurt; *Kölnische Zeitung*, Koln; *Kölnische Volkszeitung*, Köln; *Rheinisch-Westfälische Zeitung*, Essen; *Essener Volkszeitung*, Essen; *Münchener Post*, München; *Schleswig-Holsteinsche Volkszeitung*, Kiel; *Pariser Tageblatt*, Paris; *Berner Tagwacht*, Bern; *Zukunft*, Berlin; *L'Illustration*, Paris; *Die Tat*, Berlin; *Weltbuhne*, Berlin.

INDEX

INDEX

Lightning Source UK Ltd.
Milton Keynes UK
UKOW07f1814131017
310925UK00011B/1124/P